THE GENIUS ACT DECODED

Stablecoin Stability Transparency Compliance & Innovation

First Edition

ISBN: **978-1-967440-08-5**

Published by AIMQWEST CORPORATION®

AIMQWEST BOOKS®
AIMQWESTBOOKS.COM

Foreword

Six months ago, the journey of this book began—not merely with pen on paper or fingers on keyboards—but with the visionary recognition that we stood at the precipice of a profound transformation. Today, July 17, 2025, the day this book reaches your hands, marks an extraordinary convergence in history. As the Genius Act has just passed both the House and Senate, it now makes its way to the President's desk, poised to usher in a new era of financial innovation and regulatory clarity.

In the whirlwind that defines modern finance, digital assets and stablecoins have surged to the forefront, challenging traditional economic paradigms and reshaping our understanding of value itself. As stewards of innovation, it became clear to us at AIMQWEST CORPORATION that decoding the Genius Act was not just necessary—it was urgent. This legislation is more than policy—it represents an unprecedented effort to balance innovation with transparency, stability with flexibility, and economic ambition with ethical stewardship.

Within these pages, you will find a detailed, incisive analysis meticulously constructed to illuminate the Genius Act's intricate provisions. From foundational insights into digital asset governance to strategic implications for stakeholders across diverse sectors, each chapter is crafted not only to inform but to empower. We aim to equip you—policymakers, industry leaders, innovators, and regulators—with the tools necessary to navigate and leverage this landmark legislation effectively.

The Genius Act Decoded is crafted to be voraciously irresistible, inviting you to explore the depths of financial innovation, compliance intricacies, jurisdictional nuances, and ethical considerations in ways that are both engaging and accessible. Beyond regulatory interpretation, this book is a guide—a roadmap toward a future where technology and regulation no longer collide, but collaborate, propelling humanity toward unprecedented economic and social prosperity.

As you journey through this book, remember that you are not merely a passive reader but an active participant in a defining moment of history. The Genius Act is not just about the digitization of assets—it symbolizes our collective commitment to a transparent, fair, and inclusive financial future.

Today, as this bill moves closer to ratification, we stand together on the edge of tomorrow, fully prepared and informed, ready to shape the next chapter of our financial legacy.

Welcome to the future of finance.

AIMQWEST CORPORATION

TABLE OF CONTENTS

Introduction

In recent decades, technological innovation has profoundly reshaped global financial systems, introducing complexities and opportunities previously unimaginable. The emergence of digital assets, cryptocurrencies, and blockchain technology has challenged traditional regulatory paradigms, demanding rapid evolution from legislative and regulatory bodies worldwide. In response to these sweeping changes, the United States has witnessed significant legislative activity aimed at providing clear, coherent, and comprehensive frameworks for the integration of digital currencies into the existing financial infrastructure. Central to this wave of legislative innovation is the Genius Act, a groundbreaking piece of legislation designed to address critical gaps in digital asset governance, with particular emphasis on stablecoins and their underlying asset reserves.

The Genius Act is not merely a legislative development but a pivotal response to the convergence of technological advancement, economic necessity, and ethical obligation. Stablecoins, digital currencies pegged to stable assets such as fiat currency or precious metals, have emerged as crucial financial instruments capable of stabilizing volatility inherent in cryptocurrencies. Yet, their widespread adoption has been impeded by regulatory uncertainty, concerns over transparency, and apprehensions regarding systemic risk. The Genius Act aims precisely at these challenges, seeking to foster an environment where innovation thrives within well-defined regulatory boundaries, thereby balancing market freedom with investor protection, economic stability, and societal trust.

The impetus behind the Genius Act is multifaceted. First, there is a recognition of the transformative potential inherent in blockchain technology and digital currencies, which promise enhanced efficiency, inclusivity, and resilience in financial transactions. Concurrently, regulators have acknowledged the risks associated with unchecked technological experimentation, notably market manipulation, money laundering, and systemic vulnerabilities. By establishing clear legislative guidelines, the Genius Act endeavors to harness the positive potential of digital assets while instituting robust safeguards to mitigate associated risks.

Moreover, the Genius Act emerges from a nuanced understanding of contemporary economic dynamics. The global financial landscape has become increasingly interconnected, digitalized, and vulnerable to systemic shocks, as evidenced by financial crises and economic disruptions in recent memory. Traditional monetary and regulatory policies, designed for pre-digital economic models, have shown limitations when faced with innovations such as cryptocurrencies and decentralized finance (DeFi). Legislators drafting the Genius Act recognized the urgency of developing responsive regulatory frameworks capable of integrating these new financial instruments securely into mainstream economic channels.

A distinctive feature of the Genius Act is its proactive stance toward stablecoins specifically. Unlike cryptocurrencies with volatile valuations, stablecoins promise greater predictability and integration potential into everyday transactions, thereby attracting mainstream institutional interest. Nonetheless, without rigorous regulatory standards governing reserve backing, disclosure practices, and operational transparency, stablecoins risk undermining financial stability and public confidence. The Genius Act addresses this concern through detailed requirements on asset reserve management, rigorous compliance obligations, and transparent reporting standards. Such measures aim not only at protecting investors but also at preserving the integrity and resilience of the broader financial system.

This book, *"The Genius Act Decoded,"* sets out to meticulously dissect, analyze, and interpret the Genius Act, providing a clear, comprehensive understanding of its provisions, practical implications, and strategic relevance. Our goal is twofold: first, to equip policymakers, financial industry professionals, entrepreneurs, and technologists with a robust framework for understanding and applying the Genius Act; second, to stimulate critical thinking about how regulatory innovation can responsibly and ethically accommodate technological advancement.

Throughout this work, we will maintain a highly professional and analytical approach, carefully avoiding assumptions or ungrounded speculation. Each chapter provides not only rigorous theoretical analysis, but also practical guidance informed by expert insights, empirical case studies, and comparative regulatory contexts. The objective is to transform complex legislative language into actionable knowledge, enabling readers to make informed decisions in navigating the rapidly evolving financial landscape.

Readers of *"The Genius Act Decoded"* will gain an unprecedentedly clear and structured understanding of digital asset regulation, legislative strategy, and compliance mechanisms, with insights explicitly tailored to practical applications. The text systematically addresses jurisdictional differences and regulatory nuances, notably highlighting distinctions between states such as Wyoming and Texas, providing readers with a comprehensive picture of legislative choices and implications. By highlighting economic and ethical dimensions within the Act's provisions, this book ensures readers not only appreciate the technical details but also grasp the broader societal ramifications.

In subsequent chapters, we will delve deeply into the specifics of the Genius Act, examining its historical genesis, legislative intent, detailed provisions, and anticipated outcomes. Readers will find comprehensive analyses of critical components such as asset reserves, transparency requirements, jurisdictional considerations, and strategic implications for businesses seeking to leverage regulatory clarity for innovation and competitive advantage.

As we proceed through the various chapters, the text will guide readers methodically from foundational knowledge through advanced regulatory interpretation to practical strategic application. Each chapter builds upon the previous, ensuring a logical, cumulative progression of knowledge. This structure is intended to empower stakeholders across different sectors, equipping them not only with theoretical comprehension but also with practical frameworks and actionable insights necessary to navigate and leverage the Genius Act effectively.

In summary, this introduction has set forth the foundational context of the Genius Act, highlighting its significance, objectives, and unique contributions to digital finance regulation. In the subsequent sections, readers will explore in greater detail the genesis of the Genius Act, its critical components, practical implications, jurisdictional nuances, and broader ethical considerations. Each section will carefully adhere to a structured outline, ensuring clarity, comprehensiveness, and accessibility throughout.

Given the transformative potential of digital finance, it is essential to contextualize the Genius Act within broader legislative trends in global regulatory policy. Worldwide, policymakers and regulators have grappled with the rapid rise of blockchain-based financial technologies. Jurisdictions such as the European Union, Singapore, Japan, and Switzerland have enacted varied yet instructive frameworks that offer comparative benchmarks for understanding the Genius Act's unique approach. The act does not exist in a regulatory vacuum; rather, it represents a conscientious response reflective of global legislative practices, incorporating the best available lessons while simultaneously addressing distinctly American economic, political, and technological realities.

Consequently, understanding the Genius Act demands not only a national perspective but also an appreciation of international regulatory convergence and divergence. This comprehensive viewpoint allows stakeholders to recognize opportunities and challenges arising from global financial integration, cross-border transactions, and international market dynamics. Such awareness underscores the strategic importance of legislative clarity as a determinant of competitive advantage for jurisdictions aspiring to leadership in digital finance innovation.

Moreover, this text will consistently emphasize the importance of ethical considerations in financial innovation. As recent history demonstrates, rapid technological evolution without adequate regulatory oversight can result in detrimental consequences, including consumer exploitation, increased systemic risks, and potential destabilization of economic structures. The Genius Act explicitly aims to mitigate these concerns by embedding ethical oversight and robust consumer protection mechanisms within its regulatory framework. Through rigorous standards for transparency, accountability, and compliance, the Act seeks to foster not merely technological advancement but also the sustained trust of the public and market participants.

Within the pages of *"The Genius Act Decoded,"* readers will encounter nuanced discussions on the relationship between ethical governance and financial innovation. We argue that ethical considerations must form a central pillar of regulatory policy—not as constraints upon innovation, but as vital components enabling sustainable and responsible growth. Ethics, transparency, and accountability, therefore, become inseparable from the broader goals of economic efficiency, stability, and equitable access to financial services.

Another key dimension explored throughout this work involves the practical challenges and implementation complexities associated with the Genius Act. While legislative frameworks provide necessary guidance, translating regulatory theory into effective practice remains an intricate endeavor, requiring coordinated efforts among regulators, financial institutions, technologists, and policymakers. Stakeholders must navigate ambiguities, unintended consequences, compliance costs, and market adaptations to realize the full benefits envisioned by the Genius Act. This book meticulously dissects such practical challenges, providing clarity on potential hurdles, effective strategies for overcoming them, and actionable recommendations drawn from real-world case studies.

Furthermore, *"The Genius Act Decoded"* will provide clarity regarding compliance requirements, offering strategic insights for businesses, financial institutions, and technologists eager to operate within the Act's regulatory environment. Compliance is often viewed as a burden; however, through careful analysis, readers will recognize compliance as an opportunity for differentiation, market trust enhancement, and long-term competitive advantage. Properly executed, compliance demonstrates corporate responsibility, enhances brand integrity, and secures market positioning within the evolving landscape of digital assets and blockchain finance.

Importantly, the Genius Act arrives at a critical juncture in U.S. economic and technological history. Traditional financial infrastructures and legacy regulatory frameworks increasingly exhibit vulnerabilities when confronted with decentralized financial innovation. The Act addresses these systemic vulnerabilities by creating adaptable regulatory mechanisms, enabling responsiveness to future technological advancements and market developments. It represents proactive policymaking, not merely reactive regulation, positioning the United States as a global leader in responsible financial innovation.

An additional point worth underscoring involves the Genius Act's capacity for stimulating economic inclusion and accessibility. Digital financial innovations, when appropriately regulated, have demonstrated remarkable potential for expanding financial inclusion, particularly among traditionally underserved communities. By providing clear, transparent, and stable regulatory guidelines, the

Genius Act encourages institutions to explore innovative products capable of addressing longstanding barriers to financial participation, such as geographical constraints, infrastructure deficits, and socioeconomic inequalities.

In approaching this multifaceted analysis, the authors of *"The Genius Act Decoded"* have drawn upon diverse expertise in financial policy, regulatory analysis, blockchain technology, and ethical governance. The interdisciplinary approach ensures comprehensive coverage, precision, and depth in analyzing each aspect of the Genius Act. Moreover, the text incorporates extensive research, rigorous data analysis, and informed commentary from industry leaders, policymakers, and academic experts, providing readers with authoritative, credible, and balanced insights.

Throughout this work, we shall employ structured comparative analysis, detailed legislative commentary, practical case examples, and strategic foresight methodologies. The purpose of these methodological choices is clear: to empower readers with sophisticated analytical tools and actionable knowledge, enabling them to grasp fully the implications of the Genius Act and leverage opportunities arising from it. Whether readers seek legislative clarity, business insights, compliance guidance, or strategic advice, *"The Genius Act Decoded"* offers thorough, authoritative, and accessible coverage.

This introductory exploration, therefore, sets the stage for the detailed analyses contained within subsequent chapters. By firmly establishing the Genius Act's significance, intentions, challenges, and implications, readers are well-prepared to engage meaningfully with the comprehensive discussions that follow. As the chapters unfold, we shall delve systematically into the historical, regulatory, economic, jurisdictional, and ethical dimensions of this critical piece of legislation, providing a holistic view that is simultaneously practical, insightful, and forward-looking.

Finally, this book emphasizes the broader societal context in which the Genius Act operates, recognizing that financial legislation does not function in isolation but profoundly impacts public trust, market dynamics, and the global perception of financial innovation. Recent financial crises and regulatory failures underscore the crucial role that transparent, effective, and ethical legislative frameworks play in securing public confidence. The Genius Act is expressly designed to restore, preserve, and enhance public trust by embedding accountability, transparency, and reliability into the foundations of digital financial infrastructure.

Furthermore, as the global economy increasingly depends on digital systems, the Genius Act takes on added importance for safeguarding economic resilience and national security. Regulatory clarity helps mitigate systemic vulnerabilities arising from the digital interconnectivity of global financial markets, ensuring that technological innovations serve national economic interests rather than undermine them. Through clearly articulated compliance standards and robust oversight mechanisms, the Genius Act fosters an environment conducive to responsible growth, stability, and innovation.

This book also provides a roadmap for legislators and regulators, elucidating how the Genius Act can serve as a blueprint for future regulatory initiatives. It demonstrates how proactive legislative approaches can keep pace with technological developments, serving as a model for agile policymaking that anticipates rather than merely reacts to innovation. By capturing the lessons learned through drafting and implementing the Genius Act, the book offers invaluable guidance for future policy developments, ensuring continued relevance and adaptability in rapidly evolving technological landscapes.

Moreover, the Genius Act's significance extends beyond immediate economic and technological considerations—it also addresses fundamental questions about the relationship between technology, governance, and democracy. The Act reflects a commitment to democratic accountability and public oversight in financial innovation, providing mechanisms to ensure that digital asset growth aligns with broader societal interests rather than narrow corporate or speculative objectives. Thus, the Genius Act implicitly acknowledges the need for inclusive governance processes that engage diverse stakeholders in shaping the future of digital finance.

To ensure the comprehensive nature and enduring relevance of *"The Genius Act Decoded,"* each chapter systematically addresses anticipated developments and emerging trends within the regulatory environment. Readers will not only understand the Genius Act as it stands today but will also be equipped to anticipate future legislative adjustments, market adaptations, and technological advancements. This foresight will enable stakeholders to remain strategically agile and proactively engaged, adapting seamlessly to regulatory changes as they emerge.

Finally, it is critical to note that the Genius Act is ultimately an invitation to dialogue—among legislators, technologists, financial professionals, ethicists, and the broader public. Through informed, transparent, and inclusive discussions, stakeholders can collaboratively navigate the complexities of digital finance, shaping outcomes that benefit society as a whole. *"The Genius Act Decoded"* aspires to be an essential resource in fostering these dialogues, providing clarity, guidance, and informed analysis necessary for meaningful engagement.

In closing this introduction, we set forth clearly the ambitious scope and critical importance of the Genius Act, underscoring its transformative potential in reshaping the landscape of digital finance and regulation. As readers proceed through subsequent chapters, they will gain a thorough understanding of the legislative intricacies, strategic implications, practical applications, and ethical considerations encapsulated within this landmark Act. The detailed analysis that follows serves to illuminate these complexities, ensuring that readers emerge not only well-informed but also strategically empowered.

With this comprehensive context in place, we now transition to a detailed examination of the Genius Act, beginning with its historical evolution, legislative rationale, and foundational objectives.

Chapter 1: Understanding the Genius Act

Legislative History and Development

To fully appreciate the significance of the Genius Act, it is essential to examine its legislative history and the socio-economic environment that spurred its creation. The Act did not emerge spontaneously; rather, it is the product of extensive legislative deliberations, informed by economic necessity, technological innovation, and public demand for transparency and financial stability.

In the years preceding the Genius Act, digital assets and cryptocurrencies proliferated rapidly, fundamentally altering traditional financial practices. Initially, regulatory agencies responded with caution, uncertain how best to address novel challenges posed by decentralized finance (DeFi), cryptocurrency volatility, and blockchain innovations. Early attempts at regulation were fragmented, characterized by inconsistent guidance and jurisdictional ambiguity, creating confusion among market participants and stifling investment and innovation.

Recognizing these deficiencies, legislative bodies sought clearer guidelines to navigate the complexities presented by rapidly evolving digital assets. In response, preliminary discussions within congressional committees began emphasizing the need for comprehensive legislation addressing stablecoins—digital currencies pegged to stable assets, typically fiat currencies, commodities, or other reserve instruments. Stablecoins attracted significant attention due to their promise of stability, transparency, and potential for mainstream adoption compared to other more volatile cryptocurrencies.

Initially, legislative momentum gathered slowly, gaining traction as digital asset markets grew in scale and economic significance. Major incidents involving significant financial losses, fraudulent schemes, and systemic risk vulnerabilities demonstrated the urgency for coherent regulatory frameworks. Several high-profile cases underscored this necessity, including the collapse of algorithmically governed stablecoins, fraudulent initial coin offerings (ICOs), and widespread manipulation within cryptocurrency exchanges. These events galvanized policymakers, intensifying bipartisan efforts to establish regulatory standards capable of stabilizing markets, protecting investors, and encouraging responsible innovation.

This convergence of economic, political, and technological factors culminated in the introduction of the Genius Act. Drafted through a collaborative process involving legislators, regulatory agencies, industry stakeholders, and financial experts, the Act aimed explicitly at addressing key vulnerabilities exposed by previous regulatory deficiencies. Public consultations, congressional hearings, expert testimony, and rigorous debate shaped the Genius Act into comprehensive legislation balancing innovation with stringent safeguards.

A notable aspect of the Genius Act's legislative journey was its bipartisan support, reflecting the universal recognition of digital finance's importance across political divides. Throughout its drafting process, legislators consistently emphasized consumer protection, economic stability, and ethical responsibility—principles that ultimately defined the Act's central provisions. By incorporating diverse stakeholder input, the Genius Act was crafted not only to resolve immediate regulatory gaps but also to anticipate future challenges posed by technological evolution.

The final text of the Genius Act, therefore, represents a delicate balance among various competing interests: innovation versus regulation, freedom versus oversight, and agility versus stability. It recognizes digital assets' transformative potential while also mandating strict transparency standards, rigorous compliance protocols, and robust consumer protections. Moreover, the Act explicitly targets stablecoin issuance, mandating rigorous requirements for asset reserves, disclosure practices, and operational transparency—areas identified as critical vulnerabilities in earlier market disruptions.

Throughout this chapter, we will closely examine the Genius Act's primary provisions and the detailed rationale underpinning each regulatory measure. The goal is to equip readers with a thorough understanding of legislative intent, enabling clear insights into anticipated regulatory outcomes. Additionally, readers will gain perspective on how the Genius Act fits within broader U.S. financial legislation trends, highlighting its unique contributions and potential implications for future regulatory developments.

Core Provisions and Objectives

Central to the Genius Act are specific provisions designed to govern digital assets comprehensively, with an emphasis on stablecoins. The legislation includes clearly articulated definitions, categorization frameworks, asset reserve management guidelines, transparency mandates, compliance obligations, and enforcement mechanisms. Understanding these provisions thoroughly is essential for appreciating the Act's intended impact and strategic implications.

The Genius Act defines stablecoins explicitly as digital currencies designed to maintain stable value through backing by reserves consisting of traditional currencies, commodities, or other approved financial assets. This precise definition is foundational, establishing a legal basis upon which subsequent regulatory obligations are constructed. Clarity in definition is intended to eliminate ambiguity that previously hindered regulatory consistency and undermined investor confidence.

Another core objective of the Genius Act involves stringent asset reserve management protocols. Stablecoin issuers must maintain reserves equivalent to the total value of issued stablecoins, ensuring liquidity and solvency. These reserves are subject to regular audits by certified independent auditors, with audit reports mandatorily disclosed to regulatory authorities and publicly accessible. Such provisions directly address earlier criticisms about inadequate transparency and insufficient reserve backing, thus aiming to bolster investor trust and market stability.

Transparency and disclosure requirements constitute additional critical provisions within the Genius Act. Issuers are compelled to provide detailed, timely disclosures regarding their stablecoin reserves, operational policies, compliance practices, and associated risks. Regular reporting is mandated, outlining asset composition, liquidity management strategies, and operational protocols. Enhanced transparency empowers investors and regulators to evaluate issuer credibility effectively, reducing informational asymmetry and fostering healthier market dynamics.

Compliance and enforcement provisions also play pivotal roles within the Genius Act. Issuers must establish comprehensive compliance programs demonstrating adherence to the Act's rigorous standards. Regulatory authorities are empowered with broad enforcement powers, including authority for inspection, investigation, penalties, and sanctions in cases of non-compliance. The Act's enforcement framework underscores accountability, creating incentives for strict adherence while providing regulators with necessary tools to swiftly and effectively address potential violations.

Collectively, these core provisions reflect the Genius Act's overarching objectives: fostering innovation within clearly defined regulatory parameters, safeguarding market stability, protecting consumer interests, and enhancing transparency. Each provision addresses distinct regulatory gaps identified through past market failures, ensuring comprehensive coverage and robust market integrity.

In subsequent chapters, we shall explore each provision in greater depth, examining practical implications, operational challenges, strategic considerations, and compliance obligations from business and regulatory perspectives. However, at this stage, readers should clearly understand the foundational objectives and primary mechanisms the Genius Act employs to achieve its regulatory goals.

Key Stakeholders and Anticipated Impact

The Genius Act is distinguished not only by its provisions but also by the diversity and scope of stakeholders it affects. A comprehensive understanding of the Genius Act requires identifying and analyzing these key stakeholders, as well as examining the anticipated impacts and strategic implications the Act will have on their operations, opportunities, and responsibilities.

Primary stakeholders impacted directly by the Genius Act include stablecoin issuers, financial institutions, regulatory agencies, investors, consumers, and technology providers. Secondary stakeholders encompass legislators, advocacy groups, academic researchers, and international regulatory bodies monitoring developments within U.S. financial legislation. Each stakeholder group encounters unique challenges and opportunities under the Act's provisions, necessitating targeted strategic responses and adaptation processes.

Stablecoin Issuers stand at the forefront of the Genius Act's regulatory impact. These entities, ranging from innovative startups to established financial corporations, must significantly adjust operational practices to meet compliance obligations. Issuers must now maintain rigorous reserve management protocols, regular auditing procedures, transparent reporting systems, and comprehensive compliance programs. While compliance presents initial operational challenges and increased costs, it also offers substantial long-term benefits—enhanced market credibility, consumer trust, and regulatory certainty. By providing clearly defined legal standards, the Genius

Act facilitates strategic innovation, potentially accelerating stablecoin market expansion and broader adoption.

For **Financial Institutions**, the Genius Act introduces both opportunities and complexities. Banks, payment processors, and financial intermediaries previously cautious about engaging with digital assets now encounter clear regulatory guidelines, encouraging active participation. Such institutions may confidently integrate stablecoins into mainstream financial services, offering innovative products and streamlined transaction channels. Conversely, institutions must invest in compliance infrastructure, develop specialized expertise, and adjust existing risk management frameworks, initially incurring increased operational expenditures. However, the clarity provided by the Genius Act ultimately positions institutions strategically, enabling them to benefit from growing digital asset market opportunities.

Regulatory Agencies experience significant implications stemming from the Genius Act. Agencies responsible for enforcement and oversight, including the Securities and Exchange Commission (SEC), Commodity Futures Trading Commission (CFTC), and Office of the Comptroller of the Currency (OCC), are provided enhanced enforcement powers and clearer mandates, reducing jurisdictional overlap and ambiguity. While initial challenges include adapting regulatory capacities, expertise, and oversight mechanisms to effectively implement the Act, clearer statutory guidelines ultimately facilitate more efficient, targeted regulatory actions. Enhanced clarity and proactive regulation significantly reduce systemic risks, streamline enforcement processes, and strengthen investor protection efforts.

Investors and Consumers benefit considerably from the Genius Act's rigorous transparency and disclosure requirements. Prior to the Act, investors navigated digital asset markets marked by opacity, volatility, and speculative risks. Mandatory reserve disclosures, audit transparency, and stringent compliance standards empower investors with critical information necessary for informed decision-making, risk assessment, and asset allocation. Consumers, likewise, gain enhanced protections from fraudulent activities, market manipulation, and systemic vulnerabilities previously prevalent in loosely regulated environments. These enhanced protections foster greater public confidence and broader market participation, catalyzing growth and stability in digital asset markets.

For **Technology Providers** and innovators, the Genius Act creates a favorable environment characterized by regulatory certainty, clearer legal standards, and reduced market-entry barriers. Technological entrepreneurs, blockchain developers, and financial technology startups previously deterred by regulatory ambiguity can confidently develop innovative solutions aligned with clear legislative guidelines. The Genius Act incentivizes technological innovation focused explicitly on transparency, compliance efficiency, and financial security, spurring the creation of compliance solutions, reserve management technologies, and auditing platforms tailored specifically to stablecoin issuers.

Secondary stakeholders, including legislators and international regulators, closely observe the Genius Act's implementation, interpreting outcomes as potential precedents for broader regulatory frameworks. Legislators derive insights from the Act's effectiveness in balancing regulatory oversight with technological innovation, potentially replicating successful strategies in other financial and technological domains. International regulators benefit from observing U.S. regulatory experiments, integrating lessons learned into their own legislative frameworks, thereby contributing to global regulatory harmonization and collaboration.

Broader Regulatory Context and Comparative Analysis

Understanding the Genius Act further necessitates positioning it within the broader regulatory context and contrasting it with international regulatory approaches. The Act represents a distinctly American regulatory innovation, addressing specific challenges within the U.S. financial ecosystem; however, it also draws upon international precedents and contributes to global regulatory discourse.

Globally, regulatory approaches to digital assets vary widely. For example, the European Union's Markets in Crypto-assets (MiCA) regulation offers comprehensive guidelines covering digital asset issuance, trading, and custody. Singapore's Monetary Authority of Singapore (MAS) maintains clear regulatory standards emphasizing risk management, consumer protection, and market integrity. Japan and Switzerland similarly enacted proactive frameworks, balancing innovation support with strict compliance mandates. Comparative analysis highlights that while the Genius Act shares similarities with these international standards—particularly transparency and reserve management—it distinguishes itself through a more focused, stablecoin-centric regulatory approach.

Analyzing international regulatory frameworks reveals that successful digital asset regulations typically share common principles: transparency, robust compliance requirements, consumer protection, and rigorous oversight. By embedding these principles deeply into its structure, the Genius Act aligns with international regulatory best practices while maintaining sufficient flexibility to address uniquely American market conditions and systemic risks. This balance positions the United States as both a responsible participant in international financial innovation and a global leader in regulatory foresight.

Through this comparative lens, stakeholders better appreciate the Genius Act's strategic value—not merely as a domestic policy measure but as part of a coordinated global response to digital finance regulation. Businesses, investors, and regulators gain actionable insights into the Act's alignment with global regulatory trends, enabling informed decisions regarding international collaboration, market expansion, and compliance strategies.

The Genius Act also underscores the United States' commitment to leadership in global financial innovation. By proactively addressing regulatory gaps, systemic vulnerabilities, and market inefficiencies, the Act sends a powerful signal internationally: the U.S. aims to remain a premier hub for responsible innovation in digital finance. This commitment fosters international collaboration, attracts global investment, and enhances the U.S. economy's competitive positioning.

Legislative Intent and Strategic Vision

In understanding any landmark legislation, grasping the underlying intent and strategic vision of lawmakers is essential. The Genius Act is no exception. Its explicit and implicit objectives reveal strategic priorities that shape both the nature of the regulations themselves and the behaviors they aim to incentivize among market participants.

Primarily, the Genius Act articulates a clear intent to **reduce systemic financial risk**. Lawmakers, aware of historical episodes of instability triggered by inadequate oversight—such as the financial crisis of 2008, and more recently, the volatility associated with cryptocurrencies—recognized the urgency of safeguarding financial systems against potential destabilization from inadequately regulated digital assets. By introducing stringent asset reserve requirements and transparent auditing protocols, the Act strategically aims to preempt systemic vulnerabilities, providing stability not just for investors but for the broader economic ecosystem.

Another cornerstone of legislative intent lies in **protecting consumer interests and investor confidence**. Prior to the Genius Act, consumer trust in digital asset markets suffered significantly due to frequent occurrences of fraud, market manipulation, and opacity around reserve backing. Lawmakers explicitly structured the Act to rebuild consumer confidence, leveraging transparency as a tool to restore credibility. Mandating thorough, frequent, and transparent disclosure of reserve holdings and operational practices represents a clear legislative vision focused on empowering consumers with the necessary information to make informed, secure decisions.

Furthermore, legislators designed the Genius Act with the goal of fostering **financial inclusion and innovation**. Lawmakers viewed stablecoins as critical financial instruments capable of providing more inclusive, efficient, and accessible financial services, particularly to underserved populations. To achieve this inclusivity without sacrificing safety, the Act promotes innovation by delineating precise boundaries within which experimentation can safely occur. Clear regulatory frameworks reduce market-entry barriers, encouraging responsible innovation and expanding financial services to populations traditionally excluded from formal financial systems.

The strategic vision of the Genius Act also encompasses maintaining **global economic competitiveness**. Legislators understood the imperative of establishing the United States as an attractive jurisdiction for digital asset innovation, amid global competition from jurisdictions actively courting blockchain-based enterprises. Thus, the Genius Act strategically positions the U.S. as a competitive and responsible leader, balancing progressive innovation with necessary oversight. Such leadership attracts global investments and retains domestic talent, directly contributing to long-term economic resilience and growth.

Moreover, legislators crafted the Genius Act to explicitly demonstrate an ongoing commitment to **ethical financial governance**. Ethical considerations permeate the Act's core provisions, reflecting a vision of financial markets governed not solely by profit maximization, but by principles of fairness, accountability, and social responsibility. By embedding ethical oversight into legislative mandates—such as transparent reserve management and robust consumer protection mechanisms—the Act reinforces the importance of ethics as integral to sustainable financial innovation.

Anticipating Challenges and Regulatory Adaptation

Despite its careful construction, implementation of the Genius Act inevitably presents practical challenges. Legislative frameworks, regardless of their sophistication, encounter complexities during operationalization.

Stakeholders, therefore, must anticipate these challenges and proactively engage in adaptive regulatory strategies.

One significant challenge involves **regulatory enforcement capacity**. While clear statutory guidelines assist regulators, implementation often reveals capacity constraints—whether in terms of skilled personnel, technological infrastructure, or financial resources. Regulatory agencies must rapidly scale capabilities, ensuring sufficient oversight to enforce compliance without inadvertently stifling market activity through overly bureaucratic or inefficient processes.

Moreover, stakeholders must anticipate potential **ambiguities and interpretive differences** arising from the Genius Act's language. While clarity in definitions and provisions remains a priority, practical application typically surfaces nuanced interpretative challenges. Regulators and market participants must actively collaborate through consultative frameworks to clarify uncertainties, prevent unintended consequences, and ensure consistent implementation across jurisdictions.

Another critical implementation challenge pertains to the **costs associated with compliance**. Smaller enterprises, particularly innovative fintech startups, face disproportionately higher burdens from rigorous compliance requirements, potentially disadvantaging these smaller players against established institutions with deeper resources. Legislators and regulatory bodies must remain sensitive to such disparities, developing supplementary support mechanisms—such as streamlined compliance processes or transitional support measures—to ensure equitable market participation without sacrificing regulatory integrity.

Strategic Recommendations for Stakeholders

Given these anticipated challenges, stakeholders—including legislators, regulators, financial institutions, technology providers, and investors—must adopt proactive strategies to effectively operationalize the Genius Act.

Legislators and regulators are advised to maintain open and continuous **stakeholder engagement channels**. Regular consultations, forums, and feedback mechanisms will ensure responsive governance, clarify ambiguities quickly, and adapt regulatory frameworks dynamically to emerging market developments. Additionally, investments in **regulatory technology (RegTech)** can enhance oversight efficiency, reduce enforcement burdens, and streamline compliance processes for market participants.

Financial institutions and stablecoin issuers should prioritize **strategic investments in compliance infrastructure and innovation**. Rather than viewing compliance purely as a cost center, institutions are encouraged to treat compliance capabilities as strategic differentiators. Investing in sophisticated audit, transparency, and reserve management technologies will not only facilitate regulatory compliance but also enhance competitive positioning, customer trust, and operational efficiency.

Technology providers and innovators have unique opportunities to develop and market specialized solutions tailored to the Genius Act's requirements. Innovations in **audit automation, blockchain-based reserve verification, and real-time reporting technologies** represent lucrative avenues for market leadership and strategic partnerships. By proactively aligning innovation with regulatory compliance needs, technology providers can significantly capitalize on emerging market opportunities created by the Act.

Investors and consumers, meanwhile, must actively leverage newly available transparency resources for informed decision-making. Enhanced access to transparent disclosures empowers investors with comprehensive information critical for effective risk management. Investors are encouraged to prioritize engagements with compliant issuers, rewarding transparency and ethical governance through capital allocation decisions.

Conclusion and Chapter Summary

This first chapter has established a comprehensive understanding of the Genius Act, including its legislative history, core provisions, key stakeholders, strategic intent, and anticipated implementation challenges. Readers have gained critical insights into the Act's broader economic, ethical, and strategic dimensions, positioning them to navigate subsequent detailed analyses effectively.

By clearly framing the Act within its historical and legislative context, readers can appreciate the strategic foresight inherent in its provisions. Stakeholders now understand precisely how and why the Genius Act emerged, the challenges it addresses, and the opportunities it unlocks for responsible innovation within digital finance.

Moving forward, readers are well-prepared for deeper analyses of each regulatory component, detailed jurisdictional considerations, practical compliance guidance, and broader ethical and economic implications explored in subsequent chapters. This foundational understanding serves as an essential backdrop, equipping stakeholders with the analytical tools necessary for informed strategic engagement with digital finance regulation.

As the following chapters unfold, readers will further explore the Genius Act's intricate regulatory landscape, comprehensively evaluating how it reshapes the U.S. financial environment and positions America at the forefront of responsible global financial innovation.

Final Reflections on the Genius Act's Foundational Significance

In finalizing this foundational chapter, it is essential to appreciate fully the Genius Act's role as a benchmark in contemporary financial regulation. Beyond mere technical legislation, the Act signifies a progressive vision for the future, encapsulating the complex interplay between technological innovation, regulatory oversight, and economic resilience. It demonstrates a rare legislative capacity for forward-thinking, balancing immediate regulatory demands with long-term strategic considerations.

The Genius Act arrives during a critical era characterized by increasing digitalization, technological disruption, and economic interconnectedness. The future-oriented approach embedded in its provisions allows legislators, regulators, and market participants alike to adapt more confidently to ongoing and unforeseen shifts in the financial ecosystem. Rather than static, prescriptive regulation, the Act embodies a flexible yet robust legislative framework designed explicitly for adaptability in the face of continual technological evolution.

Moreover, by explicitly embedding ethical considerations and transparency mandates at its core, the Genius Act reinforces public trust, not merely in digital assets or stablecoins but broadly in financial innovation. The explicit commitment to transparency and consumer protection signifies a crucial shift toward prioritizing long-term sustainability over short-term speculative gains, a significant departure from earlier regulatory environments that prioritized growth above stability and accountability.

In conclusion, the Genius Act sets a clear precedent for future financial regulation, illustrating how proactive, balanced, and ethically grounded legislation can sustainably support economic innovation. As subsequent chapters will further illustrate, the Genius Act not only redefines the immediate landscape of stablecoin regulation but serves as an influential blueprint for broader legislative efforts addressing the complex dynamics between emerging technologies and economic stability.

This chapter has comprehensively laid the foundation, enabling stakeholders to deeply understand the Genius Act's historical emergence, regulatory objectives, stakeholder impacts, and anticipated implementation dynamics. Equipped with these insights, readers can confidently navigate the detailed analyses and practical guidance offered in the forthcoming chapters.

Chapter 2: The Financial Landscape Pre-Genius Act

Overview of Digital Asset Regulation Before the Act

To fully grasp the transformative impact of the Genius Act, it is essential to first consider the regulatory environment that existed prior to its enactment. Historically, the regulatory framework governing digital assets in the United States was characterized by considerable fragmentation, ambiguity, and reactive rather than proactive regulatory measures. This fragmentation arose from multiple regulatory bodies—including the Securities and Exchange Commission (SEC), Commodity Futures Trading Commission (CFTC), the Federal Reserve, the Office of the Comptroller of the Currency (OCC), and state-level authorities—each asserting varying degrees of jurisdiction over digital financial innovations.

Initially, regulators struggled to classify digital assets consistently, generating significant uncertainty within the financial marketplace. Cryptocurrencies and stablecoins occupied ambiguous legal territories, at times regarded as securities, commodities, or currencies, depending on the context and the regulatory agency involved. This lack of definitional clarity significantly impaired investor confidence, inhibited widespread adoption by financial institutions, and slowed technological innovation.

Moreover, early regulatory responses were primarily reactive, prompted by high-profile financial disruptions, fraud cases, and market manipulations rather than proactive policy strategies. Regulators issued warnings and enforcement actions largely on a case-by-case basis, creating a landscape marked by ad-hoc decisions rather than cohesive regulatory guidance. Such a reactive posture undermined investor trust and increased market instability, exacerbating volatility and systemic vulnerabilities.

Prior to the Genius Act, stablecoins—digital assets explicitly designed for price stability through reserve backing—were regulated inadequately. Stablecoins' potential as reliable financial instruments was undermined by insufficient oversight of reserve management, opaque operational practices, and sporadic or incomplete disclosures. Regulatory ambiguity allowed some issuers to maintain insufficient reserves, generating significant risks to investors and financial systems, especially in cases of large-scale redemptions or market stress.

The fragmented regulatory environment also created jurisdictional tensions between state and federal authorities. States like Wyoming moved proactively to foster innovation-friendly environments, establishing clear guidelines to attract digital asset businesses. Wyoming's pioneering legislative efforts, including defining stablecoins explicitly and establishing frameworks for asset custody and banking services, contrasted sharply with more cautious or restrictive approaches adopted elsewhere. Such jurisdictional disparities contributed to market confusion, regulatory arbitrage, and uneven standards across the national market.

Challenges Faced by Businesses and Regulators

The fragmented pre-Genius Act regulatory environment imposed substantial practical and strategic challenges for businesses operating in digital finance. Entrepreneurs and financial institutions faced significant uncertainties, unsure how to structure compliant operations, manage risk exposure, or predict future regulatory changes. High levels of ambiguity increased compliance costs, hindered innovation, and constrained investment in digital finance infrastructure.

Businesses seeking regulatory compliance often found themselves navigating contradictory or overlapping guidelines from multiple authorities. The lack of harmonized federal standards complicated strategic planning, increased operational complexity, and necessitated costly, customized legal and compliance frameworks for each jurisdiction. Smaller enterprises, lacking extensive legal and compliance resources, suffered disproportionately, disadvantaged relative to larger financial entities better positioned to absorb compliance-related expenditures.

Furthermore, ambiguous reserve and disclosure requirements created a challenging competitive landscape for responsible stablecoin issuers. Without clear regulatory standards, less diligent issuers frequently benefited economically by minimizing compliance investments, operating with inadequate reserves, or adopting less transparent disclosure practices. Responsible actors faced a competitive disadvantage, highlighting the urgent necessity for comprehensive regulatory clarity.

Regulators themselves faced equally daunting challenges under pre-Genius Act conditions. Fragmented jurisdictional authority, overlapping responsibilities, and ambiguous legislative frameworks constrained regulators' effectiveness. Resource limitations, insufficient technical expertise, and inadequate regulatory technologies hindered timely and effective oversight. Moreover, regulators frequently lacked clear legislative mandates defining their scope, authority, and enforcement powers concerning digital assets, further complicating their responsibilities.

Regulators were also challenged by rapid technological evolution. Innovations frequently outpaced regulatory capacity, complicating oversight and increasing systemic risks. The sheer complexity of digital asset technologies—including decentralized finance platforms, algorithmic stablecoins, and blockchain-based smart contracts—required specialized expertise, advanced technology infrastructure, and agile regulatory responses, capabilities regulators often struggled to develop quickly.

Case Study: Limitations of Existing Frameworks

To concretely illustrate pre-Genius Act regulatory limitations, we consider a high-profile case: the collapse of algorithmically governed stablecoins. Algorithmic stablecoins, unlike asset-backed stablecoins, sought stability through algorithmic adjustments of supply and demand rather than explicit reserve backing. Initially appealing for their innovative structure and lower compliance costs, these instruments faced critical vulnerabilities under volatile market conditions.

One notable collapse involved TerraUSD (UST), an algorithmic stablecoin that dramatically failed to maintain its peg during market stress. Without clear regulatory guidelines on reserve requirements or transparency obligations, UST operated with insufficient resilience mechanisms, eventually triggering a rapid loss of investor confidence and widespread market contagion. The collapse had extensive implications, causing substantial financial losses, harming public trust in digital finance broadly, and spotlighting critical gaps in regulatory oversight.

This case underscores multiple regulatory limitations prevalent pre-Genius Act:

- **Ambiguous Definitions and Classification:** Regulators lacked clear guidelines to categorize algorithmic stablecoins, complicating oversight responsibilities.
- **Inadequate Reserve Standards:** No robust reserve management or liquidity requirements existed, leaving investors exposed to substantial risks.
- **Insufficient Transparency and Disclosures:** Without mandated transparency or auditing requirements, investors had limited visibility into issuer practices or stability mechanisms.
- **Limited Regulatory Response Capacity:** Regulators responded primarily post-collapse, offering reactive enforcement rather than preventative safeguards.

Consequently, the UST collapse epitomized the systemic risks inherent in pre-Genius Act regulatory fragmentation, dramatically illustrating the urgent need for clear, proactive, and comprehensive regulatory frameworks such as those later introduced by the Genius Act.

Consequences of Regulatory Ambiguity

The regulatory ambiguity characterizing the pre-Genius Act era significantly constrained innovation and market expansion, adversely impacting investor confidence, market integrity, and the strategic development of the digital finance industry. This lack of clarity resulted in numerous adverse consequences that shaped the financial landscape, necessitating the legislative intervention ultimately embodied by the Genius Act.

Primarily, regulatory uncertainty increased **market volatility and systemic risk**. In the absence of standardized rules, digital asset markets experienced extreme fluctuations driven by speculative activity, uninformed investment decisions, and manipulative trading practices. High volatility deterred institutional investors and traditional financial institutions from meaningful engagement with cryptocurrencies and stablecoins, leaving markets dominated by speculative rather than strategic investments. Consequently, sustained market stability—critical to broader financial integration—remained elusive.

Furthermore, regulatory ambiguity contributed to **consumer vulnerabilities and investor harm**. Inconsistent disclosure standards allowed unscrupulous issuers to exploit informational asymmetries, misleading investors about asset backing, reserve adequacy, or operational risks. Cases of fraud, market manipulation, and inadequate reserves emerged repeatedly, undermining public trust. High-profile losses among retail investors, attracted by promises of stable returns, emphasized the urgent necessity of clear, enforceable regulatory standards protecting consumers from financial exploitation.

Another detrimental consequence was the phenomenon of **regulatory arbitrage**, in which businesses strategically relocated or structured operations specifically to exploit regulatory disparities between jurisdictions. Companies seeking minimal compliance burdens favored regions with less stringent regulatory requirements, undermining efforts to establish uniform standards and consistent investor protections. Regulatory arbitrage not only fragmented the market but also complicated oversight efforts, effectively diluting the potential of responsible, innovative financial technologies.

Moreover, regulatory fragmentation fostered **competitive inequities among market participants**. Larger financial institutions, equipped with sophisticated compliance infrastructure and substantial resources, navigated ambiguous regulatory environments relatively successfully. In contrast, smaller, innovative startups—lacking comparable resources—faced disproportionate compliance costs and operational complexities. This imbalance disadvantaged innovation-driven firms, limiting market competition and potentially stifling innovation precisely when digital finance offered transformative economic opportunities.

Institutional Responses and Market Adjustments

In response to these regulatory uncertainties, financial institutions, technology providers, and market participants employed various adaptive strategies, attempting to navigate and mitigate regulatory risk. Such institutional responses significantly shaped the financial landscape preceding the Genius Act and demonstrated the market's inherent adaptability, albeit under constrained conditions.

Financial institutions adopted **cautious engagement strategies**, often limiting their exposure to digital assets due to regulatory uncertainty. Major banks, investment firms, and payment processors approached digital finance conservatively, prioritizing traditional financial instruments over cryptocurrencies or stablecoins perceived as inadequately regulated. Such cautious institutional participation restricted market liquidity, prevented mainstream adoption, and slowed technological innovation due to limited institutional support and capital inflows.

Simultaneously, certain financial institutions adopted a strategy of establishing **specialized compliance departments and internal regulatory frameworks** to anticipate evolving compliance demands. Despite ambiguous external guidance, leading institutions developed sophisticated internal risk assessment models, reserve verification processes, and internal audit procedures to mitigate potential regulatory penalties. This approach, though costly, allowed institutions to strategically position themselves for future regulatory clarity, ultimately shaping the industry standards subsequently formalized by the Genius Act.

Technology companies and blockchain innovators similarly responded to regulatory ambiguity through **jurisdictional arbitrage and strategic partnerships**. Some businesses relocated operations or incorporated subsidiaries in jurisdictions offering clearer regulatory guidelines—such as Wyoming in the United States or internationally, in countries like Switzerland or Singapore. Others partnered with established financial institutions possessing advanced compliance expertise, strategically leveraging institutional credibility to reduce perceived regulatory risk.

These adaptive strategies, however innovative, also incurred significant opportunity costs. Jurisdictional arbitrage fragmented innovation clusters, dispersing talent and capital across diverse regions rather than consolidating them strategically within cohesive national or regional innovation ecosystems. Compliance-driven partnerships frequently imposed bureaucratic constraints, limiting innovative potential and agility among smaller fintech enterprises.

Economic and Strategic Costs of Pre-Act Fragmentation

Beyond immediate market responses, regulatory fragmentation preceding the Genius Act carried substantial broader economic and strategic costs. These included diminished international competitiveness, constrained innovation potential, and reduced economic efficiency.

One significant cost was the erosion of the United States' **global competitive position in digital finance innovation**. Countries adopting proactive regulatory frameworks attracted substantial international investment and entrepreneurial talent, potentially placing the U.S. at a strategic disadvantage. Financial innovators seeking predictable regulatory environments increasingly favored jurisdictions outside the U.S., reducing domestic innovation, employment opportunities, and capital inflows crucial for national economic growth.

The ambiguous regulatory landscape also resulted in substantial **economic inefficiencies**. Regulatory uncertainty elevated compliance costs across financial institutions and technology providers, diverting resources away from innovation toward legal, compliance, and administrative expenditures. Smaller firms disproportionately bore these compliance burdens, diverting resources away from critical investments in technological innovation, strategic growth, and market expansion.

Moreover, fragmented regulatory environments adversely impacted the broader strategic development of the digital asset ecosystem by creating barriers to **collaboration and interoperability**. Financial institutions and technology providers developed isolated compliance systems, technology platforms, and operational practices tailored to divergent regulatory interpretations. This isolation inhibited interoperability across systems, limiting network effects and hindering broader ecosystem efficiencies critical for digital finance's long-term economic integration.

Finally, the fragmented regulatory environment impaired the potential of digital finance to facilitate **financial inclusion**. Stablecoins and blockchain-based financial services initially promised significant opportunities to extend financial services to underserved communities by reducing transaction costs, improving access, and increasing financial participation. However, regulatory ambiguity constrained such inclusive innovations by increasing perceived risk, limiting institutional participation, and constraining widespread market adoption. Consequently, digital finance's broader social and economic potential remained significantly under-realized, underscoring the urgency of coherent regulatory reform embodied subsequently by the Genius Act.

Lessons Learned from Early Regulatory Approaches

Analyzing the fragmented regulatory environment preceding the Genius Act provides critical insights and valuable lessons that have shaped contemporary approaches to digital asset governance. Identifying these lessons clearly elucidates why comprehensive legislation became a strategic imperative and underscores the Genius Act's eventual significance in rectifying systemic regulatory weaknesses.

First, a key lesson derived from early regulatory experiences involves the necessity of **definitional clarity and consistency**. Regulators learned that ambiguities in the classification of digital assets directly contributed to market confusion, regulatory loopholes, and enforcement difficulties. Early approaches that failed to clearly distinguish between cryptocurrencies, stablecoins, commodities, and securities created operational uncertainty and deterred institutional engagement. Thus, the Genius Act explicitly incorporated precise definitions and standardized classifications, aiming to eliminate such ambiguities permanently.

Second, regulators learned the importance of adopting a **proactive rather than reactive regulatory stance**. Previous reliance on reactive measures—such as enforcement actions initiated after-market failures—proved ineffective in preventing systemic crises, investor harm, or fraudulent activities. This reactive stance often resulted in costly post-event interventions rather than preventive measures. Consequently, the Genius Act embodies proactive regulatory principles, explicitly designed to preempt systemic vulnerabilities through stringent reserve standards, transparency mandates, and robust enforcement frameworks.

Third, historical regulatory experiences underscored the critical need for **cross-agency coordination and jurisdictional harmonization**. Prior regulatory fragmentation demonstrated the practical limitations of isolated regulatory responses by multiple agencies with overlapping or unclear jurisdictions. Regulatory silos hindered effective oversight, compliance clarity, and enforcement efficiency. Recognizing this limitation, legislators drafting the Genius Act prioritized clear agency responsibilities, streamlined enforcement coordination, and enhanced inter-agency collaboration, significantly reducing jurisdictional ambiguities.

Fourth, early regulatory approaches revealed the necessity of **regulatory agility and adaptability**. Regulatory frameworks rigidly defined in traditional financial paradigms proved ineffective in addressing rapid technological innovation and evolving market practices. Such rigidity led to mismatches between regulatory expectations and technological realities, frequently resulting in regulatory obsolescence and diminished market relevance. Hence, the Genius Act strategically integrates flexible provisions designed to accommodate technological advancements, ensuring sustained regulatory relevance amid ongoing innovation.

Finally, previous regulatory efforts demonstrated the critical importance of **transparency and consumer protection** as foundational elements of sustainable market development. Markets lacking robust disclosure requirements, auditing standards, or transparency mandates repeatedly demonstrated susceptibility to fraud, systemic vulnerabilities, and investor exploitation. In response, the Genius Act embedded transparency and consumer protection centrally within its regulatory architecture, explicitly requiring reserve disclosures, operational transparency, and rigorous compliance standards.

Pre-Act Regulatory Milestones and Influential Developments

Despite its fragmented and reactive nature, the pre-Genius Act period was characterized by significant regulatory milestones and influential developments that helped pave the way for comprehensive legislation. Highlighting these milestones contextualizes the Genius Act as a logical culmination of earlier regulatory efforts rather than an isolated legislative event.

A notable regulatory milestone was the **2019 SEC Framework for "Investment Contract" Analysis of Digital Assets**, which provided initial clarity by outlining criteria for classifying digital assets as securities under the Howey Test. Although limited, this framework established foundational principles, guiding issuers and market participants regarding securities regulations applicable to digital assets. Its introduction represented a crucial first step toward regulatory standardization, signaling regulatory intent to clarify digital asset classifications.

Another influential development involved the **OCC's 2020 Interpretive Letter**, authorizing national banks to provide cryptocurrency custody services. This milestone significantly enhanced institutional confidence by explicitly recognizing cryptocurrency custodianship as a legitimate banking service. It represented a strategic regulatory acknowledgment of digital assets' growing economic importance, facilitating institutional participation by removing critical operational uncertainties.

Similarly impactful was Wyoming's **Digital Asset Regulatory Framework**, initiated in 2019, positioning Wyoming as a pioneering jurisdiction through comprehensive digital asset legislation. Wyoming explicitly classified digital assets, including stablecoins, established regulatory clarity on custody and reserve requirements, and created innovative banking frameworks specifically for digital asset businesses. Wyoming's proactive legislation provided influential regulatory precedents, significantly shaping later federal legislative approaches embodied in the Genius Act.

Internationally, the **European Union's Markets in Crypto-assets (MiCA) regulation** and Switzerland's progressive blockchain legislation provided influential comparative frameworks, demonstrating successful approaches to comprehensive digital asset governance. These international milestones not only shaped domestic legislative thinking but also underscored the urgency of establishing robust, competitive U.S. regulatory frameworks to remain globally relevant and strategically positioned in digital finance innovation.

Building Momentum for the Genius Act

The gradual accumulation of regulatory challenges, lessons learned, and influential milestones gradually built momentum for comprehensive legislative intervention embodied by the Genius Act. By the early 2020s, consensus emerged across political, regulatory, and market communities that existing fragmented approaches were insufficient, necessitating clear, standardized, and proactive legislative frameworks.

This building momentum was significantly accelerated by heightened public and political attention following high-profile market disruptions, fraudulent activities, and consumer harms associated with inadequate digital asset regulation. Public and investor demands for regulatory clarity intensified, creating political impetus for comprehensive legislative intervention. Additionally, bipartisan recognition of digital finance's economic potential created legislative consensus around the necessity of proactive regulation balancing innovation, consumer protection, and systemic stability.

Simultaneously, extensive stakeholder consultations, industry advocacy, and regulatory deliberations increasingly emphasized the urgent necessity of federal legislation addressing digital asset markets comprehensively. Industry leaders, regulators, and market participants repeatedly advocated clearer definitions, reserve requirements, transparency mandates, and streamlined regulatory oversight, creating critical advocacy coalitions supporting the Genius Act's legislative development.

Furthermore, international regulatory developments highlighted earlier underscored urgency in establishing competitive U.S. regulatory frameworks to retain global leadership in digital finance innovation. Recognizing international regulatory benchmarks and comparative advantages, lawmakers strategically prioritized the Genius Act to secure U.S. economic competitiveness and prevent further migration of digital finance innovation abroad.

Thus, by the mid-2020s, conditions were ripe for comprehensive legislation, resulting in the drafting, debate, and eventual enactment of the Genius Act. The Act represented not merely reactive legislation but a proactive, strategic, and deliberate policy response to accumulated regulatory challenges, lessons learned, and stakeholder advocacy developed over preceding years.

Conclusion and Chapter Summary

The fragmented, ambiguous, and reactive regulatory environment that characterized the pre-Genius Act period posed substantial barriers to innovation, investor confidence, and economic stability in digital finance. This chapter has thoroughly examined the regulatory landscape preceding the Genius Act, highlighting critical challenges faced by market participants, regulators, and policymakers alike.

By identifying key regulatory deficiencies, including ambiguous definitions, inadequate transparency standards, reactive enforcement approaches, jurisdictional fragmentation, and competitive inequities, this chapter clearly demonstrates why comprehensive legislative intervention became necessary. Additionally, influential regulatory milestones and proactive jurisdictional efforts—such as Wyoming's pioneering legislation and international regulatory advancements—underscored the urgent necessity and strategic benefits of coherent, comprehensive digital asset regulation.

The cumulative lessons learned from early regulatory experiences ultimately informed the Genius Act's legislative architecture, ensuring definitional clarity, proactive oversight, transparency mandates, enhanced consumer protection, and harmonized inter-agency coordination. As subsequent chapters explore, the Genius Act effectively addresses each critical deficiency highlighted in this chapter, representing a logical and strategic legislative evolution rather than an isolated event.

With a solid understanding of the regulatory environment preceding the Genius Act, readers are now fully prepared to engage deeply with subsequent analyses exploring the Genius Act's specific provisions, jurisdictional considerations, practical implications, and broader strategic significance within the contemporary financial landscape.

Chapter 3: Decoding the Key Provisions

Introduction to Core Regulatory Components

Having thoroughly reviewed the financial landscape preceding the Genius Act, we now shift our focus to a comprehensive exploration of the Genius Act's primary provisions. The Genius Act distinguishes itself through precise and extensive guidelines explicitly addressing stablecoins, emphasizing asset reserve requirements, transparency obligations, and stringent compliance standards. Understanding these detailed provisions thoroughly equips stakeholders with the clarity necessary for strategic alignment, informed compliance, and effective participation in the evolving digital finance ecosystem.

Stablecoin Definitions and Classifications

A foundational achievement of the Genius Act lies in its explicit definition and classification of stablecoins. Prior regulatory ambiguity severely undermined investor confidence and regulatory efficiency, rendering clear legislative definitions crucial. The Genius Act addresses this by providing precise, standardized definitions:

- **Stablecoin Definition:** Under the Genius Act, stablecoins are explicitly defined as digital assets designed to maintain stable value through backing by reserves consisting of fiat currencies, commodities, government bonds, or other approved financial instruments. The legislation distinguishes stablecoins explicitly from other digital assets, such as volatile cryptocurrencies, ensuring clear regulatory differentiation.

- **Asset-Backed vs. Algorithmic Stablecoins:** Further classification distinguishes between asset-backed stablecoins—those directly backed by tangible, audited reserves—and algorithmic stablecoins, whose value relies on supply-demand mechanisms and algorithmic protocols. The Genius Act subjects asset-backed stablecoins to stringent reserve verification and disclosure mandates, reflecting their explicit reliance on reserve stability. Algorithmic stablecoins face rigorous compliance requirements around operational transparency and risk disclosures, acknowledging their unique stability mechanisms and inherent vulnerabilities.
- **Systemically Significant Stablecoins:** Additionally, the Genius Act introduces the classification of "systemically significant stablecoins," defined as those with substantial market capitalization or extensive usage capable of materially impacting broader economic stability. Such stablecoins are subject to enhanced regulatory oversight, stringent reserve audits, and higher transparency obligations, reflecting their greater systemic influence.

These clear classifications eliminate prior ambiguities, providing stakeholders with precise legal frameworks essential for compliance planning, investor confidence, and market transparency. By explicitly differentiating stablecoins based on structural attributes and systemic significance, the Genius Act strategically tailors regulatory standards to specific asset characteristics and associated risks.

Asset Reserve Requirements

A central pillar of the Genius Act involves rigorous requirements for stablecoin asset reserves, directly addressing earlier vulnerabilities stemming from insufficiently backed issuances. Prior regulatory frameworks failed to establish clear reserve obligations, resulting in significant investor exposure to liquidity risks and issuer insolvency.

The Genius Act establishes stringent reserve management protocols, including:

- **Full Reserve Backing:** Stablecoin issuers are mandated to maintain reserves equivalent in value to 100% of their issued stablecoins. This requirement ensures immediate liquidity, preserves issuer solvency, and prevents systemic risk stemming from potential reserve shortfalls during periods of market volatility or redemption pressures.
- **Qualifying Reserve Assets:** The Act precisely specifies allowable reserve asset classes, including fiat currencies (USD, EUR, GBP, etc.), short-term government bonds, highly liquid securities, and precious metals (gold, silver, etc.). Each asset class is subject to stringent liquidity and creditworthiness criteria designed to minimize market risk and ensure rapid liquidation if required.
- **Custodial Standards:** Reserve assets must be held securely by qualified custodians—regulated

financial institutions explicitly approved to provide custody services. Custodians must meet rigorous security, operational, and reporting standards, ensuring asset safety, transparency, and regulatory oversight.

- **Frequent Auditing and Verification:** Stablecoin issuers must engage independent, certified auditors to conduct regular reserve audits, verifying compliance with mandated reserve backing and custodial standards. Audit results must be publicly disclosed promptly, enhancing transparency and reinforcing issuer credibility among investors and regulators.

These reserve requirements significantly elevate transparency and financial security, addressing prior vulnerabilities that undermined market confidence. By mandating rigorous reserve management practices, the Genius Act provides investors, regulators, and market participants with tangible assurances of issuer stability, solvency, and liquidity, substantially reducing systemic risks associated with stablecoin issuances.

Transparency and Disclosure Standards

One of the Genius Act's defining features lies in its rigorous transparency and disclosure requirements. Historically, insufficient transparency posed substantial risks, resulting in investor skepticism, heightened volatility, and systemic vulnerabilities. By clearly mandating transparency at multiple levels, the Genius Act explicitly aims to address these critical gaps, thereby stabilizing markets and bolstering investor confidence.

Under the Genius Act, stablecoin issuers must adhere to extensive transparency protocols, which include:

- **Reserve Transparency:** Issuers must provide detailed, regularly updated reports outlining reserve asset composition, custody arrangements, and liquidity conditions. These reports, released monthly, must clearly disclose the types of reserve assets held, their current market value, custodial institutions involved, and audit confirmations. Comprehensive disclosures equip investors with critical insights necessary for informed risk assessment and asset allocation decisions.
- **Operational Transparency:** Issuers are required to maintain publicly accessible documentation clearly describing operational procedures, redemption policies, pricing mechanisms, and governance structures. Detailed operational disclosures must clearly outline how stablecoins maintain their pegged value, redemption processes for stablecoin holders, and mechanisms employed to mitigate market volatility or redemption pressures.
- **Risk Disclosure Requirements:** Issuers must transparently communicate potential risks associated with their stablecoins, including liquidity risks, custodial risks, market volatility exposures, and technological vulnerabilities. Clear, comprehensive risk disclosures empower investors with realistic expectations regarding potential performance, contributing significantly to informed decision-making and effective risk management.

- **Third-Party Audits and Certifications:** Independent third-party auditors must verify compliance with transparency mandates, reserve adequacy, and operational procedures. Audit reports must be publicly disclosed within predefined periods after each audit, ensuring market accountability and issuer credibility. Auditors must be certified by relevant regulatory authorities, adhering strictly to prescribed auditing standards and professional conduct guidelines.

These transparency measures represent a strategic regulatory intervention designed explicitly to enhance accountability, reduce informational asymmetry, and stabilize investor expectations. By mandating comprehensive transparency protocols, the Genius Act significantly reduces risks previously inherent within digital asset markets, directly contributing to improved financial stability and market efficiency.

Compliance and Enforcement Mechanisms

Rigorous transparency and reserve management standards within the Genius Act are further reinforced by robust compliance and enforcement provisions. Effective regulatory enforcement requires not only clearly articulated standards but also comprehensive compliance frameworks and strong enforcement authorities capable of ensuring adherence and swiftly addressing non-compliance.

The Genius Act explicitly establishes extensive compliance obligations and enforcement powers, including:

- **Mandatory Compliance Programs:** Stablecoin issuers are required to establish comprehensive compliance programs aligned explicitly with Genius Act mandates. These programs must include internal control mechanisms, clearly defined compliance policies, regular staff training, and established reporting channels for regulatory interactions. Compliance programs must be regularly updated and independently reviewed to ensure effectiveness and alignment with evolving regulatory standards.
- **Regulatory Inspection and Investigation Powers:** Regulatory authorities—including the SEC, CFTC, and other relevant bodies—are empowered with expansive inspection, examination, and investigation capabilities. Regulators may conduct regular on-site inspections, audits, and examinations to verify compliance directly, gaining immediate access to issuer records, reserve custody facilities, and operational processes. Investigatory powers enable prompt regulatory responses to compliance concerns, significantly enhancing enforcement efficacy.
- **Reporting and Record-Keeping Requirements:** Issuers must maintain extensive, detailed records documenting compliance with reserve management standards, transparency disclosures, and operational procedures. These records must be retained securely, remain accessible for regulatory inspections, and be maintained for specified minimum periods, ensuring accountability and facilitating regulatory oversight.

- **Penalty and Enforcement Measures:** Non-compliance carries clearly defined penalties and enforcement consequences under the Genius Act. Potential regulatory actions include substantial financial penalties, issuance restrictions, operational limitations, and potential criminal referrals for severe violations involving fraud or market manipulation. Clearly articulated enforcement consequences significantly incentivize issuer compliance, deter malfeasance, and provide regulatory authorities with powerful tools to swiftly rectify non-compliance.

Collectively, these compliance and enforcement provisions demonstrate a comprehensive regulatory commitment to accountability, market integrity, and systemic stability. Effective enforcement ensures that transparency mandates and reserve management requirements achieve their intended impacts, significantly reducing systemic vulnerabilities historically characteristic of insufficiently regulated stablecoin markets.

Governance and Accountability Structures

Beyond rigorous compliance and enforcement mechanisms, the Genius Act incorporates explicit governance and accountability structures designed to foster ethical, responsible, and transparent management practices among stablecoin issuers. Effective governance is central to maintaining operational integrity, safeguarding investor interests, and ensuring ongoing regulatory compliance.

Key governance provisions under the Genius Act include:

- **Board Accountability:** Issuers must establish clearly defined governance frameworks overseen by accountable boards of directors or equivalent governing bodies. Boards must explicitly assume responsibility for ensuring compliance with reserve management obligations, transparency requirements, and internal compliance programs. Directors are held personally accountable for oversight failures or compliance breaches resulting from inadequate governance practices, incentivizing rigorous oversight and responsible management conduct.
- **Internal Audit Functions:** Issuers must implement robust internal audit functions tasked explicitly with monitoring compliance, evaluating operational procedures, and identifying potential risks or compliance gaps. Internal auditors must operate independently from management, report directly to governance boards, and maintain full authority to investigate internal operations thoroughly. Regular internal audit reports must document findings, propose corrective actions, and be readily available for regulatory inspection, significantly enhancing operational transparency and accountability.
- **Stakeholder Communication and Reporting:** Governance structures must include established processes for transparent, regular communication with stakeholders—including investors, regulatory authorities, and the broader market community. Issuers are mandated to maintain open reporting channels, providing frequent updates on operational practices, financial health, reserve conditions, and compliance statuses. Transparent

stakeholder communication reinforces trust, facilitates informed decision-making, and provides market accountability beyond mandatory disclosure requirements.

- **Whistleblower Protections:** Recognizing the critical role whistleblowers play in identifying non-compliance or fraudulent practices, the Genius Act explicitly mandates comprehensive whistleblower protection frameworks within issuer organizations. Employees or external stakeholders reporting concerns are safeguarded from retaliatory actions, provided anonymous reporting mechanisms, and guaranteed regulatory protection, significantly encouraging early detection and prevention of regulatory violations or misconduct.

Collectively, these governance and accountability structures significantly strengthen issuer responsibility, operational transparency, and regulatory compliance. Robust governance frameworks provide critical internal checks and balances, complementing external regulatory enforcement efforts and collectively ensuring sustainable market integrity.

Jurisdictional Clarification and Federal-State Coordination

An additional critical provision of the Genius Act addresses jurisdictional clarity and enhanced coordination between federal and state regulatory authorities. Historically, fragmented jurisdictional responsibilities undermined regulatory effectiveness, creating significant market uncertainty and compliance complexity. By clearly delineating jurisdictional roles and responsibilities, the

Genius Act strategically streamlines oversight, reducing regulatory overlap and ambiguity.

Key jurisdictional provisions within the Genius Act encompass:

- **Federal Regulatory Primacy:** The Act explicitly establishes federal regulatory primacy over stablecoin issuance and governance, designating specific federal agencies, notably the SEC and CFTC, as primary oversight authorities. Centralizing regulatory responsibilities enhances oversight clarity, ensures consistent enforcement standards, and prevents jurisdictional arbitrage resulting from fragmented state regulations.
- **Role of State Regulatory Authorities:** Recognizing states' historical role and valuable expertise in regulating financial services, the Genius Act maintains specific roles for state-level regulatory authorities. States retain responsibilities for licensing stablecoin businesses operating within their jurisdictions, consumer protection oversight, and enforcement of state-specific operational standards, provided these standards align explicitly with federal guidelines. State regulators also collaborate closely with federal authorities, providing local insights, investigative support, and jurisdictional enforcement assistance.
- **Interagency Coordination Framework:** The Genius Act mandates explicit coordination frameworks facilitating efficient communication and collaboration among federal and state regulatory bodies. Formalized interagency committees, regular coordination meetings, and clearly defined

information-sharing protocols significantly enhance enforcement efficacy, reduce jurisdictional conflicts, and ensure consistent regulatory interpretations and applications.

These jurisdictional clarity and coordination measures fundamentally resolve previous regulatory fragmentation challenges. By explicitly defining regulatory responsibilities, establishing robust coordination frameworks, and harmonizing enforcement standards, the Genius Act significantly reduces market uncertainty, streamlines compliance processes, and enhances regulatory effectiveness nationwide.

Operational Risk Management Standards

Recognizing that operational risks represent significant vulnerabilities in digital finance, the Genius Act explicitly mandates robust operational risk management frameworks among stablecoin issuers. Operational risks—encompassing technological failures, cybersecurity threats, and procedural vulnerabilities—can severely disrupt markets and damage investor confidence. By setting explicit operational standards, the Genius Act directly addresses these critical risks, enhancing market resilience and issuer reliability.

Key operational risk management standards under the Genius Act include:

- **Technology and Infrastructure Requirements:** Issuers must maintain resilient, secure, and scalable technological infrastructure capable of handling market stress, operational disruptions, and cyber threats. Infrastructure requirements include redundancy mechanisms, disaster recovery procedures, and real-time monitoring capabilities ensuring operational continuity under various scenarios. Robust infrastructure standards ensure reliable stablecoin performance, minimize systemic risks, and enhance investor confidence in issuer operational resilience.

- **Cybersecurity and Information Protection:** Issuers must implement comprehensive cybersecurity frameworks addressing prevention, detection, and response to cyber threats. Mandated cybersecurity practices include regular vulnerability assessments, penetration testing, multi-layered security architectures, and secure data management protocols. Issuers must also establish formal incident response plans detailing rapid threat identification, containment strategies, and timely reporting to regulatory authorities, significantly reducing cyber risk impacts on issuer operations and investor assets.
- **Business Continuity Planning (BCP):** Stablecoin issuers are required to develop, maintain, and regularly test comprehensive business continuity plans. BCP frameworks must explicitly outline contingency measures for various operational disruptions, including technological failures, natural disasters, and liquidity crises. Mandatory periodic BCP testing and reviews ensure preparedness for operational disruptions, reinforcing issuer reliability and market stability.
- **Outsourcing and Third-Party Risk Management:** Recognizing issuer reliance on third-party service providers—such as custodians, technology

vendors, and audit firms—the Genius Act mandates stringent third-party risk management frameworks. Issuers must conduct thorough due diligence, maintain contractual agreements defining clear performance standards, and regularly monitor third-party compliance with operational standards. These provisions ensure issuer accountability extends explicitly to outsourced functions, preventing operational vulnerabilities originating externally.

Collectively, these operational risk management standards significantly reduce issuer vulnerabilities, reinforcing operational resilience, investor protection, and market stability. By mandating explicit operational frameworks, the Genius Act addresses systemic risks historically characteristic of digital finance, positioning stablecoin markets for sustainable, secure growth.

Consumer Protection Measures

A cornerstone of the Genius Act involves explicit consumer protection measures designed to safeguard stablecoin investors and market participants. Historical experiences revealed extensive investor vulnerabilities arising from opaque operations, inadequate reserve disclosures, and insufficient regulatory oversight. To rectify these vulnerabilities, the Genius Act incorporates rigorous consumer protection frameworks ensuring investor confidence and market integrity.

Explicit consumer protection measures mandated by the Genius Act include:

- **Redemption Guarantees and Liquidity Access:** Issuers must provide clear redemption guarantees, explicitly outlining stablecoin redemption processes, timelines, and reserve access. Mandatory liquidity access provisions ensure investors retain immediate redemption capabilities, significantly reducing investor liquidity risks previously prevalent in less regulated environments. Issuers failing to provide timely redemptions face stringent penalties, incentivizing robust liquidity management practices.

- **Transparent Fee Structures and Pricing:** Stablecoin issuers must disclose clearly all fees, charges, and pricing structures associated with stablecoin issuance, redemption, and usage. Transparent fee disclosures prevent hidden costs, foster investor trust, and enhance informed consumer decisions, thereby contributing significantly to market fairness and investor satisfaction.
- **Consumer Complaint and Resolution Processes:** Issuers must maintain clearly defined, accessible consumer complaint mechanisms enabling investors to report concerns or disputes promptly. Complaint resolution processes must specify clear timelines, impartial investigation procedures, and transparent reporting of outcomes to regulatory authorities. Effective consumer complaint frameworks ensure timely dispute resolution, protect investor interests, and provide critical regulatory oversight insights.
- **Investor Education and Risk Communication:** Recognizing the complexity of digital finance, issuers must proactively engage in investor education initiatives clearly communicating

stablecoin risks, operational procedures, and market dynamics. Mandatory investor disclosures, educational resources, and transparent risk communications empower investors with necessary knowledge, significantly reducing informational asymmetries and enhancing market stability through informed investment decisions.

These comprehensive consumer protection measures represent critical regulatory interventions explicitly addressing historical investor vulnerabilities. By mandating rigorous transparency, liquidity guarantees, and dispute resolution frameworks, the Genius Act establishes robust investor safeguards significantly enhancing market confidence and consumer trust.

Regulatory Reporting Obligations

Integral to the Genius Act's effectiveness are its explicit regulatory reporting obligations. Clear, consistent, and timely reporting ensures ongoing transparency, enables efficient regulatory oversight, and reinforces issuer accountability. Historically, inadequate reporting requirements allowed operational opacity, hindered regulatory effectiveness, and exposed markets to systemic vulnerabilities. By clearly mandating rigorous reporting standards, the Genius Act addresses these shortcomings decisively.

Stablecoin issuers must comply with comprehensive regulatory reporting obligations, including:

- **Periodic Financial Reports:** Issuers must submit detailed, standardized financial reports to designated regulatory authorities on a quarterly basis. Reports must clearly outline asset reserve holdings, balance sheets, income statements, liquidity positions, and risk management assessments. Standardized financial reporting enables regulators to consistently evaluate issuer financial health, monitor systemic risks, and ensure compliance with mandated reserve requirements.

- **Event-Driven Reporting:** Issuers are mandated to provide immediate regulatory notifications regarding material events impacting operational stability, reserve adequacy, or market dynamics. Events triggering mandatory reports include substantial reserve fluctuations, cybersecurity breaches, technological disruptions, significant redemption demands, or regulatory investigations. Prompt event-driven reporting enables rapid regulatory responses, mitigates systemic risks, and ensures timely public transparency.
- **Annual Independent Audit Reports:** Stablecoin issuers must submit comprehensive annual audit reports verifying compliance with reserve requirements, transparency standards, and operational procedures. Audits must be conducted by certified independent auditors approved explicitly by regulatory authorities. Audit reports must detail compliance confirmations, identified risks, corrective actions, and auditor recommendations. Annual audit reports reinforce market accountability, enhance investor confidence, and ensure sustained regulatory oversight effectiveness.
- **Compliance Certification Reports:** Issuers must provide annual compliance certifications explicitly

affirming adherence to Genius Act mandates. Compliance reports must detail internal control procedures, compliance audits, identified deficiencies, and corrective actions taken. Mandatory compliance certifications ensure continuous regulatory accountability, reinforce issuer responsibility, and facilitate prompt identification and rectification of compliance gaps.

These comprehensive reporting obligations significantly enhance transparency, market stability, and regulatory effectiveness. Clear, consistent reporting requirements reduce operational opacity, empower informed investor decisions, and provide regulators with essential information necessary for proactive oversight and systemic risk management.

Capital Adequacy and Financial Stability Standards

Further augmenting market resilience, the Genius Act incorporates explicit capital adequacy and financial stability standards for stablecoin issuers. Historical experiences demonstrated substantial vulnerabilities arising from insufficient capital buffers, liquidity inadequacies, and inadequate risk management practices. By mandating rigorous capital and stability frameworks, the Genius Act proactively mitigates these vulnerabilities, significantly enhancing market resilience.

Explicit capital adequacy and financial stability standards under the Genius Act include:

- **Minimum Capital Requirements:** Stablecoin issuers must maintain explicit minimum capital levels proportionate to total stablecoin issuance and operational scale. Mandated capital buffers ensure issuer solvency, reduce systemic risks during market disruptions, and guarantee sufficient liquidity to meet redemption obligations. Issuers falling below prescribed capital thresholds must promptly restore capital adequacy or face regulatory intervention.

- **Liquidity Management Requirements:** Issuers must implement comprehensive liquidity management frameworks explicitly addressing liquidity risks, redemption obligations, and market volatility scenarios. Mandatory liquidity standards require issuers to maintain sufficient high-quality liquid assets to fulfill stablecoin redemption demands rapidly under various market conditions. Rigorous liquidity management practices ensure issuer reliability, protect investor interests, and safeguard broader financial stability.
- **Stress Testing Protocols:** Stablecoin issuers are required to conduct regular stress tests evaluating capital adequacy, liquidity resilience, and operational stability under adverse market scenarios. Mandated stress testing assesses issuer preparedness for potential crises, identifies operational vulnerabilities, and enables proactive risk mitigation strategies. Stress test results must be reported regularly to regulatory authorities, reinforcing issuer accountability and facilitating regulatory oversight effectiveness.
- **Recovery and Resolution Planning:** Recognizing issuer insolvency risks, the Genius Act mandates explicit recovery and resolution plans detailing procedures for managing operational disruptions,

liquidity shortfalls, or issuer insolvencies. Resolution frameworks must clearly outline mechanisms protecting investor assets, facilitating orderly market exit strategies, and minimizing systemic disruptions. Mandatory recovery and resolution planning significantly enhances market preparedness, ensures investor protection, and stabilizes financial markets under adverse conditions.

Collectively, these capital adequacy and financial stability standards reinforce issuer resilience, operational preparedness, and market stability. Rigorous capital buffers, liquidity management frameworks, and proactive stress testing practices significantly reduce systemic vulnerabilities, mitigate issuer insolvency risks, and position digital asset markets for sustainable, resilient growth.

Anti-Money Laundering (AML) and Countering the Financing of Terrorism (CFT) Frameworks

Recognizing the inherent risks digital finance poses regarding illicit financial activities, the Genius Act explicitly mandates comprehensive Anti-Money Laundering (AML) and Countering the Financing of Terrorism (CFT) frameworks for stablecoin issuers. Digital assets, including stablecoins, have historically faced criticism for facilitating money laundering, terrorist financing, and other illicit transactions due to insufficient regulatory oversight. To address these vulnerabilities proactively, the Genius Act requires stringent AML/CFT standards, significantly enhancing market integrity, investor security, and regulatory compliance.

Stablecoin issuers must comply explicitly with AML and CFT obligations, including:

- **Customer Due Diligence (CDD) and Know Your Customer (KYC):** Issuers must implement rigorous customer identification protocols verifying customer identities, transaction purposes, and sources of funds. Comprehensive KYC practices involve gathering detailed customer information, continuously monitoring transactional patterns, and promptly identifying suspicious activities. Effective CDD procedures significantly reduce illicit transaction risks, safeguard investor interests, and align issuer operations with global AML/CFT standards.
- **Suspicious Activity Reporting (SAR):** Stablecoin issuers are mandated to establish comprehensive suspicious activity reporting mechanisms explicitly identifying, documenting, and reporting potentially illicit transactions or customer behaviors. Issuers must file timely, detailed SARs with relevant regulatory authorities, clearly outlining suspicious activities, involved parties, transaction details, and mitigating actions undertaken. Mandatory SAR reporting significantly enhances regulatory enforcement capabilities, disrupts illicit financing activities, and reinforces market accountability.
- **Transaction Monitoring and Record-Keeping:** Issuers must maintain advanced transaction monitoring systems detecting unusual transaction patterns, anomalous account behaviors, and potential AML/CFT violations. Automated monitoring platforms must promptly identify

suspicious activities, trigger necessary investigations, and generate comprehensive audit trails. Detailed transaction records must be securely maintained and readily accessible for regulatory inspections, ensuring accountability and facilitating effective enforcement actions.

- **AML/CFT Training and Compliance Culture:** Recognizing the importance of organizational compliance cultures, the Genius Act explicitly requires stablecoin issuers to provide comprehensive, regular AML/CFT training programs for all relevant employees. Training initiatives must emphasize regulatory obligations, illicit finance risks, suspicious activity identification, and reporting responsibilities. Mandatory compliance training significantly enhances organizational AML/CFT awareness, reduces operational vulnerabilities, and fosters proactive compliance cultures.

Collectively, these comprehensive AML/CFT frameworks represent critical regulatory interventions explicitly addressing illicit finance vulnerabilities historically characteristic of digital asset markets. By mandating rigorous customer identification, transaction monitoring, and compliance training standards, the Genius Act significantly mitigates AML/CFT risks, safeguards investor assets, and positions stablecoin markets for sustainable, compliant growth.

Data Privacy and Protection Standards

As stablecoin issuance inherently involves handling sensitive consumer and financial data, the Genius Act explicitly mandates rigorous data privacy and protection standards. Historical experiences demonstrated significant risks associated with insufficient data security measures, cybersecurity breaches, and unauthorized disclosures of sensitive financial information. Recognizing these vulnerabilities, the Genius Act mandates comprehensive data privacy frameworks explicitly safeguarding consumer data and reinforcing operational integrity.

Stablecoin issuers must comply explicitly with stringent data privacy and protection obligations, including:

- **Comprehensive Data Security Protocols:** Issuers must implement robust data security measures explicitly protecting sensitive consumer information, transactional records, and financial data. Mandatory security protocols include encryption, secure storage solutions, multi-factor authentication, and advanced access controls restricting data accessibility exclusively to authorized personnel. Comprehensive data security practices significantly reduce cybersecurity risks, protect investor confidentiality, and enhance issuer credibility.
- **Incident Reporting and Data Breach Notification:** Issuers must establish clearly defined data breach notification procedures explicitly addressing cybersecurity incidents, unauthorized data disclosures, and operational disruptions compromising data security. Mandatory notification frameworks require immediate regulatory reporting following identified breaches, detailing incident scope, affected data categories, mitigation measures, and corrective actions undertaken. Prompt incident reporting significantly enhances

regulatory oversight, enables rapid mitigation responses, and reinforces issuer accountability.

- **Consumer Consent and Data Usage Transparency:** Stablecoin issuers must maintain explicit consumer consent mechanisms clearly defining permissible data usage practices, consumer rights regarding personal data, and transparent disclosures regarding data management procedures. Mandatory consent and transparency provisions empower consumers with control over data privacy, foster investor trust, and ensure issuer accountability concerning consumer data practices.
- **Regular Data Protection Audits:** Issuers must conduct periodic data protection audits explicitly evaluating data security measures, privacy compliance, and operational vulnerabilities. Independent audits must verify adherence to mandated data privacy standards, identify potential security gaps, and recommend corrective actions explicitly addressing identified vulnerabilities. Regular data protection audits reinforce operational transparency, enhance regulatory accountability, and significantly reduce issuer data security risks.

These comprehensive data privacy and protection standards represent critical regulatory interventions explicitly safeguarding consumer interests, operational integrity, and market credibility. By mandating stringent data security, transparent consumer consent mechanisms, and rigorous incident reporting procedures, the Genius Act significantly reduces data-related vulnerabilities historically characteristic of digital finance operations.

Market Surveillance and Monitoring Provisions

Acknowledging digital finance markets' inherent susceptibility to manipulation and volatility, the Genius Act explicitly incorporates robust market surveillance and monitoring provisions. Effective surveillance is crucial to identify market abuses, mitigate manipulation risks, ensure issuer compliance, and maintain overall market integrity. By mandating stringent surveillance protocols, the Genius Act directly addresses systemic vulnerabilities historically prevalent within digital asset markets.

Stablecoin issuers, as well as associated exchanges and custodians, must comply explicitly with detailed surveillance and monitoring obligations, including:

- **Real-Time Market Surveillance:** Issuers and associated market operators must implement sophisticated surveillance systems continuously monitoring stablecoin trading activities, transactional volumes, and pricing behaviors in real-time. Advanced analytics platforms must detect anomalous market movements, suspicious trading patterns, and potential manipulation activities promptly, triggering immediate investigative and regulatory reporting actions as warranted.
- **Manipulation Detection and Reporting:** Comprehensive detection systems explicitly target manipulative practices such as wash trading, spoofing, pump-and-dump schemes, and coordinated market abuse activities. Stablecoin issuers and market operators must promptly report identified manipulative behaviors, documenting detailed transactional records, involved market participants, suspicious activity characteristics, and

corrective actions undertaken to regulatory authorities.

- **Market Integrity Reporting:** Issuers must regularly submit detailed market integrity reports explicitly documenting surveillance activities, manipulation detection incidents, systemic risks identified, and market stability assessments. Mandatory integrity reporting ensures continuous regulatory oversight, reinforces market transparency, and enhances issuer accountability concerning market surveillance practices.
- **Coordination with Regulatory Authorities:** Issuers and market operators must maintain explicit coordination mechanisms with regulatory agencies responsible for market oversight, enabling efficient communication, timely reporting, and effective regulatory responses to market surveillance findings. Formalized coordination frameworks significantly enhance enforcement effectiveness, streamline compliance interactions, and ensure comprehensive regulatory surveillance coverage.

These stringent market surveillance provisions significantly reinforce market stability, transparency, and regulatory accountability. Advanced monitoring and real-time detection frameworks proactively identify potential market abuses, deter manipulative activities, and empower effective regulatory enforcement actions, substantially reducing systemic risks historically characteristic of digital asset markets.

Consumer Redress and Compensation Mechanisms

To comprehensively safeguard consumer interests, the Genius Act explicitly mandates clear, effective consumer redress and compensation mechanisms addressing potential financial losses, issuer failures, or regulatory violations. Historical experiences demonstrated significant consumer vulnerabilities in digital asset markets due to insufficient redress procedures, limited compensation frameworks, and inadequate protections against issuer insolvencies or operational failures. Recognizing these vulnerabilities, the Genius Act explicitly establishes robust consumer compensation structures significantly enhancing investor confidence, market credibility, and financial security.

Stablecoin issuers must comply explicitly with comprehensive consumer redress obligations, including:

- **Mandatory Compensation Funds:** Issuers must maintain dedicated compensation funds explicitly addressing investor losses resulting from issuer insolvency, operational failures, fraudulent activities, or regulatory violations. Compensation funds must be adequately capitalized, independently audited, and transparently managed, providing explicit consumer assurances concerning potential financial recoveries following adverse events.
- **Clear Consumer Redress Procedures:** Issuers must establish clearly defined, accessible consumer redress procedures explicitly detailing claims submission processes, eligibility criteria, evaluation methodologies, and compensation timelines. Transparent redress frameworks significantly enhance consumer trust, enable

effective dispute resolutions, and provide regulatory authorities with explicit accountability mechanisms addressing issuer misconduct or failures.

- **Regulatory Oversight of Compensation Practices:** Regulatory authorities retain explicit oversight responsibilities concerning issuer compensation funds and redress procedures. Regulators regularly evaluate compensation fund adequacy, claims management practices, and overall consumer redress effectiveness, ensuring sustained issuer compliance and consumer protection.
- **Public Transparency and Reporting:** Issuers must provide regular public disclosures explicitly documenting compensation fund statuses, consumer claims management activities, aggregate compensation amounts disbursed, and remaining fund balances. Mandatory transparency and reporting significantly enhance consumer confidence, reinforce issuer accountability, and facilitate ongoing regulatory oversight.

Collectively, these explicit consumer redress and compensation provisions represent critical regulatory interventions explicitly addressing historical consumer vulnerabilities. By mandating robust compensation frameworks, transparent redress procedures, and rigorous regulatory oversight, the Genius Act significantly reinforces consumer protection, market integrity, and financial stability.

Crisis Management and Emergency Response Protocols

Recognizing the significant market disruption potential inherent in operational failures, issuer insolvencies, or systemic crises, the Genius Act explicitly mandates rigorous crisis management and emergency response protocols. Historically, digital asset markets demonstrated notable vulnerabilities during market stress, issuer insolvencies, or technological disruptions, often exacerbated by inadequate response mechanisms. To address these vulnerabilities proactively, the Genius Act introduces detailed crisis response frameworks ensuring market stability, investor protection, and regulatory preparedness.

Stablecoin issuers must comply explicitly with comprehensive crisis management protocols, including:

- **Explicit Crisis Response Plans:** Issuers must develop, maintain, and regularly test detailed crisis response plans explicitly outlining emergency management procedures, operational continuity strategies, and market stabilization actions. Plans must detail clear roles, responsibilities, escalation procedures, and coordination mechanisms explicitly designed to manage crises swiftly and effectively.
- **Liquidity Crisis Management:** Recognizing stablecoins' inherent reliance on reserve liquidity, issuers must establish explicit liquidity crisis management procedures addressing reserve shortfalls, redemption pressures, or rapid market volatility scenarios. Liquidity management frameworks must explicitly define reserve access protocols, collateral liquidation procedures, redemption prioritization strategies, and regulatory notification obligations during liquidity crises.

- **Issuer Insolvency and Resolution Frameworks:** Explicit issuer insolvency response procedures must clearly define mechanisms protecting investor assets, orderly market exit strategies, and regulatory oversight protocols during issuer insolvencies. Resolution frameworks mandate rapid asset transfers, operational continuity provisions, and investor compensation fund activations explicitly mitigating investor losses and systemic risks during insolvency scenarios.
- **Regulatory Coordination and Notification:** Issuers must maintain explicit regulatory notification frameworks explicitly triggering immediate communications during crises, emergencies, or operational disruptions. Prompt regulatory notifications significantly enhance regulatory oversight effectiveness, facilitate rapid coordinated responses, and minimize systemic disruptions during emergencies.

Collectively, these explicit crisis management provisions significantly enhance market resilience, investor protections, and regulatory preparedness. Rigorous emergency response frameworks explicitly address systemic risks, issuer vulnerabilities, and operational disruptions historically prevalent within digital finance markets, substantially reinforcing market stability and operational continuity.

International Coordination and Cross-Border Regulatory Cooperation

Given digital finance markets' inherently global nature, effective regulatory oversight necessitates explicit international coordination frameworks and cross-border cooperation mechanisms. Historically, fragmented international regulations enabled regulatory arbitrage, cross-border illicit activities, and inconsistent investor protections across jurisdictions. Recognizing these challenges, the Genius Act explicitly mandates robust international cooperation provisions explicitly designed to harmonize global regulatory standards, enhance enforcement effectiveness, and ensure comprehensive investor protections.

Key international coordination and cooperation provisions under the Genius Act include:

- **Cross-Border Regulatory Dialogue:** Regulatory authorities overseeing stablecoin issuers must explicitly engage in regular international regulatory dialogues, harmonizing stablecoin definitions, compliance standards, reserve requirements, and transparency obligations across jurisdictions. Continuous international dialogue significantly reduces regulatory arbitrage, enhances global market stability, and facilitates effective enforcement cooperation internationally.
- **Mutual Recognition Agreements:** The Genius Act explicitly authorizes regulatory authorities to establish mutual recognition agreements (MRAs) explicitly harmonizing compliance standards, enforcement cooperation frameworks, and regulatory oversight responsibilities internationally. MRAs facilitate streamlined cross-border compliance processes, reduce operational complexities for international issuers, and ensure consistent investor protections globally.

- **International Enforcement Cooperation:** Regulatory authorities retain explicit mandates explicitly facilitating international enforcement cooperation concerning cross-border stablecoin issuances, market manipulation investigations, AML/CFT enforcement actions, and issuer misconduct cases. Formalized enforcement cooperation mechanisms significantly enhance global market integrity, reduce cross-border illicit finance risks, and reinforce consistent regulatory oversight internationally.
- **Information-Sharing Protocols:** Regulatory authorities must establish explicit international information-sharing protocols explicitly facilitating efficient cross-border regulatory communications, coordinated investigations, and mutual enforcement support. Effective information-sharing significantly enhances regulatory responsiveness, enforcement efficiency, and global systemic risk management capabilities.

These explicit international cooperation provisions significantly reinforce global market stability, investor protections, and regulatory effectiveness. Robust international coordination frameworks explicitly address cross-border regulatory challenges historically characteristic of digital asset markets, significantly reducing systemic vulnerabilities, regulatory arbitrage, and market manipulation risks internationally.

Environmental, Social, and Governance (ESG) Standards

Acknowledging the growing importance of sustainability and corporate responsibility in financial markets, the

Genius Act explicitly integrates Environmental, Social, and Governance (ESG) standards into stablecoin regulatory frameworks. Historically, digital asset markets received scrutiny regarding energy consumption, governance transparency, and social responsibility, prompting regulators to proactively address these concerns. By explicitly mandating ESG standards, the Genius Act positions stablecoin markets to align strategically with broader societal expectations regarding sustainability and responsible corporate behavior.

Key ESG compliance requirements under the Genius Act include:

- **Environmental Impact Assessments:** Stablecoin issuers must perform explicit environmental impact assessments detailing energy consumption, carbon footprints, sustainability initiatives, and strategies for reducing environmental impacts. Regular environmental reporting, transparent disclosures regarding sustainability performance, and proactive mitigation strategies significantly enhance issuer accountability concerning environmental stewardship.
- **Social Responsibility Practices:** Issuers must explicitly implement robust social responsibility frameworks addressing community engagement, consumer fairness, inclusive financial access, and ethical market practices. Mandatory social responsibility disclosures clearly communicate issuer commitments to responsible innovation, financial inclusion, and equitable consumer practices, reinforcing market integrity and investor trust.

- **Governance Transparency:** Reinforcing prior governance mandates, ESG provisions explicitly require comprehensive governance transparency concerning board diversity, ethical oversight mechanisms, and executive accountability structures. Clear governance disclosures significantly enhance organizational transparency, reinforce corporate responsibility cultures, and align issuer operations explicitly with broader ESG expectations.
- **Regular ESG Reporting:** Stablecoin issuers must regularly submit detailed ESG reports explicitly documenting environmental sustainability initiatives, social responsibility frameworks, governance transparency practices, and overall ESG performance metrics. Comprehensive ESG reporting significantly enhances market transparency, investor confidence, and issuer accountability concerning sustainable, responsible corporate behavior.

Collectively, these explicit ESG standards significantly position stablecoin markets for sustainable, responsible growth explicitly aligned with broader societal expectations regarding sustainability, ethical innovation, and corporate accountability.

Chapter Summary and Conclusion

This chapter has comprehensively decoded the Genius Act's key provisions, providing readers with explicit insights into regulatory standards, compliance obligations, and enforcement mechanisms introduced explicitly by this landmark legislation. By systematically addressing stablecoin definitions, asset reserve requirements, transparency mandates, compliance standards, governance frameworks, operational risk management, consumer protections, AML/CFT protocols, data privacy measures, market surveillance practices, crisis management protocols, international coordination frameworks, and ESG standards, the Genius Act comprehensively addresses digital asset market vulnerabilities historically undermining market integrity, investor confidence, and systemic stability.

Each regulatory provision explicitly targets prior market challenges, strategically positions stablecoin issuers for responsible innovation, and significantly enhances regulatory oversight effectiveness. Explicit compliance mandates and rigorous enforcement authorities explicitly ensure continuous issuer accountability, operational transparency, and investor protections, significantly reinforcing digital asset markets' integrity, resilience, and stability.

As subsequent chapters explore practical implementation considerations, jurisdictional analyses, and strategic implications for stakeholders, this foundational understanding of Genius Act provisions explicitly equips readers with necessary regulatory clarity and compliance insights critical for informed strategic decisions, effective regulatory alignment, and proactive market engagement.

Chapter 4: Practical Implications for Businesses

Navigating the Genius Act's Business Environment

The Genius Act represents more than mere regulatory reform—it constitutes a transformative shift in the financial services landscape, creating **both** new opportunities and operational complexities for businesses. Successfully navigating this legislative framework demands meticulous planning, rigorous adherence to compliance requirements, and strategic foresight. Businesses ranging from fintech startups to established financial institutions will benefit greatly from a clear, step-by-step understanding of how to operationalize compliance and leverage regulatory clarity to competitive advantage. This chapter provides comprehensive, actionable guidance on implementing stablecoin issuance procedures, conducting thorough cost analyses, overcoming operational challenges, and adopting regulatory best practices.

Strategic Approach to Stablecoin Issuance under the Genius Act

Businesses considering stablecoin issuance under the Genius Act must approach the process strategically, guided by an informed understanding of the legislative environment. The initial stage involves clearly defining the stablecoin's intended use case, the targeted customer base, and the appropriate asset reserve structure. The Genius Act differentiates between asset-backed and algorithmically managed stablecoins, and choosing the appropriate category is foundational to subsequent compliance obligations.

For asset-backed stablecoins, businesses must prioritize detailed reserve management strategies. Clear documentation regarding the type, composition, liquidity, and custody arrangements of reserve assets is mandatory. Algorithmic stablecoins must instead articulate precise and transparent stabilization mechanisms, explicitly addressing the operational resilience and reliability of these mechanisms under diverse market scenarios.

Businesses must also evaluate jurisdictional considerations early in the issuance planning process, accounting for both federal and state-level regulatory variations. Engaging with specialized legal counsel early to outline jurisdictional compliance strategies will streamline subsequent regulatory engagements. This strategic foresight mitigates potential regulatory friction and accelerates the issuance process.

Detailed Issuance Process: Step-by-step Guidance

To ensure regulatory compliance and operational efficacy, stablecoin issuers must adhere to a systematic, structured issuance process under the Genius Act. This step-by-step approach offers clarity, minimizes oversight risks, and aligns with regulatory expectations.

Step 1: Comprehensive Market and Regulatory Analysis

Businesses must begin by conducting thorough market and regulatory analyses, identifying consumer demands, competitive positioning, and precise regulatory requirements. A clear understanding of the Genius Act's compliance obligations—including asset reserve standards, transparency mandates, and operational risk management guidelines—is foundational. This stage also includes strategic benchmarking against leading stablecoin issuers to identify best practices and compliance standards.

Step 2: Legal and Governance Framework Establishment
Issuers must develop a detailed legal and governance framework, establishing clear organizational responsibilities for regulatory compliance, internal controls, and operational oversight. Forming a governance board or compliance committee accountable for regulatory adherence significantly enhances organizational alignment and accountability. It is critical to appoint a compliance officer with explicit regulatory experience to manage direct communication with regulatory authorities and oversee internal compliance practices.

Step 3: Custodial Arrangements and Reserve Asset Management
Stablecoin issuers must establish formal partnerships with qualified custodians explicitly responsible for reserve management and security. Rigorous custodial due diligence—including evaluations of custodial financial stability, compliance practices, security protocols, and disaster recovery measures—is necessary. Explicit custodial agreements must define clear responsibilities, asset access protocols, and transparency obligations, reinforcing issuer accountability and market trust.

Step 4: Liquidity Management Planning
Adequate liquidity management plans are essential to maintain issuer solvency, manage redemption demands effectively, and minimize market disruptions. Businesses must implement robust liquidity frameworks, conducting regular liquidity stress tests under diverse market conditions. Explicit plans detailing reserve liquidation strategies, redemption prioritization processes, and contingency funding measures are critical for compliance and operational resilience.

Step 5: Obtaining Regulatory Approvals and Licensing
Issuers must secure explicit licensing approvals from relevant federal and state regulatory authorities. The approval process typically requires comprehensive documentation, clearly detailing reserve adequacy, custodial arrangements, operational practices, compliance measures, and governance structures. Engaging proactively with regulatory authorities, providing timely documentation, and transparently addressing regulatory inquiries significantly streamlines approval processes, enabling rapid market entry.

Step 6: Issuance, Transparency, and Market Communication
Upon receiving regulatory approvals, issuers may commence stablecoin issuance. Implementing explicit transparency measures—including regular public disclosures of reserve composition, independent audit results, operational updates, and compliance certifications—is critical. Establishing dedicated communication channels with market participants, investors, and regulatory bodies significantly enhances market credibility, investor confidence, and regulatory compliance effectiveness.

Cost Considerations and Financial Analysis

Issuing stablecoins under the Genius Act involves significant financial considerations, extending beyond reserve asset allocations to encompass broader operational, compliance, and administrative costs. Businesses must carefully assess these financial implications to ensure sustainable operational feasibility and competitive market positioning.

Initial Reserve Allocation

The primary financial requirement involves allocating adequate reserves to back stablecoin issuance fully. The Genius Act mandates full reserve backing, necessitating careful planning of reserve asset composition, liquidity, and custody arrangements. Depending on issuance scale, initial reserve allocations may range from modest sums suitable for smaller issuers to substantial financial commitments necessary for large-scale issuances. Accurate financial planning and strategic reserve management are essential to maintain ongoing regulatory compliance.

Custodial Fees and Asset Management Costs

Custodial arrangements involve regular fees typically ranging between 0.5% to 1.5% annually of total assets under management. Additionally, asset management activities—including reserve asset purchases, ongoing monitoring, liquidity management, and regular audits—further contribute to annual operational costs. Businesses must strategically evaluate custodial services, seeking optimal balances between security, service quality, and cost-efficiency.

Compliance Infrastructure Expenses
Comprehensive compliance infrastructures are essential under the Genius Act, incurring significant financial outlays. Costs associated with internal compliance staffing, outsourced compliance advisory services, compliance training, regulatory reporting, and compliance technology solutions typically represent substantial annual expenses. Smaller issuers might expect annual compliance costs between \$5,000 to \$15,000, while larger entities should budget accordingly higher amounts reflective of greater operational complexity.

Auditing and Transparency Costs
Regular independent auditing and mandatory transparency disclosures impose further ongoing expenses. Annual independent audits mandated by the Genius Act typically range between \$3,000 and \$10,000, depending on issuer complexity, audit scope, and auditor reputation. Monthly transparency disclosures require internal administrative resources dedicated explicitly to documentation, record-keeping, and investor communications, cumulatively adding annual costs of approximately \$2,000 to \$5,000.

Operational and Administrative Expenses
Operational overhead—including administrative staffing, technological infrastructure, cybersecurity measures, internal audit functions, risk management protocols, and operational continuity planning—also represents significant ongoing costs. These expenses vary widely based on issuer scale, operational complexity, and technological sophistication, typically adding an annual expense ranging between \$10,000 and \$50,000, proportionate to business size and market ambitions.

Cumulative Financial Outlook

Summarizing these various financial elements, stablecoin issuance under the Genius Act typically involves substantial annual financial commitments. For illustrative purposes, businesses issuing stablecoins with modest reserve backing (e.g., $50,000 to $100,000) might anticipate annual costs ranging between $20,000 to $60,000. Larger issuers—especially those managing substantial reserve assets exceeding $1 million—must strategically plan for proportionally higher annual expenditures reflective of heightened regulatory scrutiny and operational complexity.

While the Genius Act provides much-needed clarity, it also introduces several operational challenges that businesses must overcome to achieve sustained compliance and operational effectiveness. Recognizing these potential hurdles and proactively adopting targeted solutions ensures resilience, market credibility, and long-term viability in digital finance markets.

Challenge 1: Complex Compliance Management

One of the most pressing challenges for stablecoin issuers under the Genius Act involves managing the complexities of compliance. The act mandates strict reserve management, transparency disclosures, operational audits, and regular reporting, which can be burdensome, particularly for newer, smaller businesses.

Solution: Deploy Advanced Regulatory Technologies (RegTech)

Businesses should leverage advanced regulatory technologies designed specifically to streamline complex compliance tasks. Regulatory technology platforms can automate reserve verifications, compliance tracking, and reporting obligations, significantly reducing manual workload, increasing accuracy, and ensuring timely regulatory adherence. Investing early in RegTech platforms fosters operational efficiency and helps maintain compliance accuracy, freeing internal resources to focus on strategic business development.

Challenge 2: Rigorous Liquidity and Reserve Management

The Genius Act's stringent requirements for full reserve backing and rigorous liquidity management create additional pressures on issuers. Businesses must effectively forecast redemption requests, maintain high liquidity, and ensure continuous adherence to regulatory standards.

Solution: Adopt Robust Liquidity Management Protocols

Companies should develop comprehensive liquidity frameworks, including detailed scenario planning and stress-testing methodologies. Establishing explicit liquidity risk management protocols, emergency contingency funding sources, and detailed asset liquidation procedures ensures preparedness during heightened market volatility or unexpected redemption events. Partnering with liquidity management experts and financial advisors provides issuers with critical external insights, reinforcing their overall risk management capabilities.

Challenge 3: Maintaining Cybersecurity and Data Integrity

As stablecoins rely heavily on digital platforms and distributed ledger technology, cybersecurity threats and data integrity risks become paramount concerns under the

Genius Act's stringent operational risk management guidelines.

Solution: Implement Multi-Layered Cybersecurity Defenses
To proactively mitigate cybersecurity threats, businesses should adopt comprehensive, multi-layered cybersecurity frameworks. This includes regular penetration testing, vulnerability assessments, data encryption, robust access controls, and incident response planning. Engaging cybersecurity specialists for regular security audits and obtaining cybersecurity insurance further enhances preparedness against potential threats and aligns with regulatory standards for data protection and operational resilience.

Challenge 4: Third-party Risk Management
Stablecoin issuers frequently rely on third-party service providers—such as custodians, auditors, technology vendors, and compliance consultants—each potentially introducing additional risks related to non-compliance, service disruptions, or data breaches.

Solution: Strengthen Third-party Oversight and Due Diligence
Robust third-party risk management begins with rigorous due diligence processes during vendor selection, including comprehensive assessments of service providers' compliance capabilities, security protocols, and operational resilience. Explicit contractual agreements clearly defining compliance responsibilities, service-level expectations, data handling requirements, and penalties for non-compliance significantly mitigate third-party risks. Regular performance monitoring, periodic reviews, and

comprehensive audits further reinforce third-party accountability and operational stability.

Challenge 5: Transparent Market Communication
Under the Genius Act, maintaining consistent, clear, and transparent communication with regulators, investors, and market participants can pose significant operational burdens, especially for businesses with limited experience in investor relations or public communications.

Solution: Establish Dedicated Investor Relations and Transparency Channels
Businesses should implement dedicated transparency and investor communication channels, including regularly updated investor portals, detailed operational reports, and clear public disclosure practices. Allocating internal resources or employing external communication specialists to manage investor relations activities ensures clarity, consistency, and regulatory compliance. Effective market communication fosters trust, enhances issuer credibility, and ensures compliance with transparency obligations outlined in the Genius Act.

Challenge 6: Regulatory Coordination Across Jurisdictions
Given the dual federal and state regulatory frameworks emphasized by the Genius Act, stablecoin issuers often encounter jurisdictional complexity that may delay licensing processes or cause unexpected compliance hurdles.

Solution: Early Engagement and Legal Expertise
Proactively engaging experienced legal counsel to navigate jurisdictional complexities and establishing early communication with relevant regulatory authorities at both state and federal levels significantly reduces jurisdictional challenges. Formulating detailed jurisdictional compliance

strategies in advance and maintaining ongoing dialogue with regulators enable businesses to anticipate potential regulatory changes and respond efficiently, facilitating smoother licensing processes and enhanced operational agility.

Compliance under the Genius Act is not merely about avoiding regulatory penalties; it represents a strategic opportunity for businesses to differentiate themselves competitively. Adopting best practices beyond minimum compliance standards positions businesses favorably within the marketplace, fostering investor trust, regulatory credibility, and operational excellence.

Best Practice 1: Cultivating an Organizational Culture of Compliance
Businesses that prioritize a culture of proactive compliance establish clear competitive advantages. Ensuring that compliance principles are deeply embedded in the organization's values and operations significantly enhances regulatory adherence and mitigates risks of inadvertent breaches. Company leadership should visibly endorse and communicate compliance priorities, provide regular compliance training across all organizational levels, and integrate compliance metrics within overall business performance evaluations.

Best Practice 2: Regular Internal Audits and Self-Assessments
Conducting routine internal audits and systematic self-assessments is integral to sustaining ongoing compliance with Genius Act requirements. Regular internal audits proactively identify compliance gaps, operational inefficiencies, or inaccuracies in reserve management, allowing businesses to implement timely corrective

measures. Independent internal audit functions that report directly to governance boards or compliance committees ensure objectivity, transparency, and ongoing accountability, thus significantly strengthening regulatory adherence.

Best Practice 3: Leveraging Regulatory Technology
Adopting advanced regulatory technologies to automate compliance processes is another pivotal best practice. RegTech platforms facilitate seamless reserve asset tracking, real-time compliance reporting, automated audit trails, and streamlined AML/KYC monitoring. Businesses that integrate advanced compliance technology solutions reduce operational overhead, minimize human error, enhance accuracy, and ensure timely regulatory submissions, positioning themselves effectively for sustained compliance.

Best Practice 4: Proactive Regulatory Engagement
Actively engaging with regulatory authorities through consistent communication and transparent reporting fosters goodwill, enhances trust, and improves regulatory relations. Businesses should proactively seek guidance, clarify regulatory expectations, and communicate openly regarding compliance challenges or operational developments. Establishing dedicated liaison roles, such as compliance officers tasked explicitly with regulatory interactions, facilitates efficient communication and positions issuers advantageously when navigating potential compliance challenges.

Best Practice 5: Advanced Liquidity and Reserve Management
Implementing sophisticated liquidity and reserve management practices is essential for sustained regulatory

compliance under the Genius Act. Businesses should conduct detailed liquidity forecasting, perform scenario-based stress testing, and maintain diversified, highly liquid reserve assets. Advanced reserve management strategies ensure ongoing issuer solvency, provide reassurance to investors, and mitigate potential redemption pressures or market volatility impacts.

Best Practice 6: Enhanced Cybersecurity and Data Protection
Given the digital nature of stablecoin technologies, prioritizing robust cybersecurity frameworks constitutes a critical compliance best practice. Businesses should employ multi-layered cybersecurity strategies, including secure encryption protocols, comprehensive access controls, real-time threat detection systems, and frequent vulnerability assessments. Maintaining comprehensive cybersecurity insurance coverage and regularly consulting with external cybersecurity experts further reinforce data integrity, regulatory compliance, and operational resilience.

Best Practice 7: Effective Crisis Management and Contingency Planning
Proactive crisis management and comprehensive contingency planning significantly enhance issuer resilience and regulatory compliance capabilities. Businesses should develop explicit crisis response plans, clearly defining emergency management roles, response strategies, communication protocols, and recovery processes. Conducting regular crisis management drills and scenario planning exercises ensures preparedness for operational disruptions, issuer insolvencies, or unexpected liquidity crises, thereby strengthening market trust and regulatory confidence.

Best Practice 8: Continuous Compliance Training and Education
Regular compliance training and education programs reinforce organizational awareness and adherence to Genius Act obligations. Issuers should conduct periodic training sessions focused explicitly on regulatory updates, reserve management standards, operational risk management protocols, cybersecurity best practices, and AML/CFT compliance requirements. Sustained education efforts enhance internal compliance awareness, reduce inadvertent compliance breaches, and reinforce organizational accountability.

Best Practice 9: Transparent Communication and Stakeholder Engagement
Proactively maintaining open, clear, and consistent communication with stakeholders—investors, regulators, market participants, and internal teams—is essential for sustained compliance and market credibility. Issuers should implement comprehensive stakeholder communication plans, providing timely updates on reserve statuses, audit findings, compliance certifications, operational developments, and strategic objectives. Effective stakeholder engagement fosters trust, enhances issuer reputation, and aligns strategically with the Genius Act's transparency mandates.

Best Practice 10: Integrating Environmental, Social, and Governance (ESG) Standards
Aligning issuer operations with rigorous ESG standards presents another strategic best practice under the Genius Act. Businesses should perform explicit environmental impact assessments, clearly articulate social responsibility practices, and transparently communicate governance frameworks. Regular ESG performance disclosures

demonstrate issuer commitment to responsible, sustainable business practices, significantly enhancing market reputation, investor attractiveness, and regulatory alignment.

Conclusion: Positioning Businesses for Long-term Success

Effectively navigating the practical implications of the Genius Act requires more than mere regulatory compliance. Successful stablecoin issuers proactively embrace operational best practices, comprehensive risk management frameworks, and transparent market communications. By systematically addressing compliance complexities, adopting advanced technological solutions, maintaining proactive regulatory engagement, and prioritizing robust operational planning, businesses significantly enhance their competitive positioning and market resilience.

The strategic implications of adhering to the Genius Act extend beyond regulatory requirements—businesses embracing a proactive, rigorous compliance ethos are positioned not only for immediate market entry but also for sustainable long-term growth. As the digital financial landscape continues to evolve, stablecoin issuers that proactively leverage regulatory clarity provided by the Genius Act will be strategically positioned at the forefront of financial innovation, economic inclusion, and market leadership.

Chapter 5: Jurisdictional Analysis: Wyoming vs. Texas

Wyoming's Pioneering Role in Digital Asset Legislation

In the rapidly evolving regulatory landscape of digital assets, Wyoming has distinguished itself as a beacon of innovation and forward-thinking governance. The state has strategically positioned itself as a pioneer in digital asset legislation, enacting comprehensive laws designed to encourage blockchain innovation, stablecoin issuance, and digital asset custody, thus offering a compelling blueprint for jurisdictions worldwide.

Wyoming’s approach to digital asset regulation has been marked by its proactive stance, driven by a clear legislative vision aimed at economic diversification, attracting innovative businesses, and positioning itself at the forefront of digital finance. The state’s legislators recognized early that traditional financial regulations, often characterized by ambiguity and reactionary enforcement, were inadequate to govern emerging technologies such as blockchain, cryptocurrencies, and stablecoins. Consequently, Wyoming pursued a series of legislative initiatives meticulously crafted to provide regulatory clarity, legal certainty, and operational flexibility for digital asset businesses.

The hallmark of Wyoming's regulatory innovation is the Special Purpose Depository Institution (SPDI) charter, introduced in 2019. SPDIs are specialized banking institutions designed to handle digital assets, stablecoins, and cryptocurrencies. They offer custody, fiduciary, and transactional services tailored specifically to blockchain-based businesses, thereby providing a critical bridge between traditional financial systems and the emerging digital economy. Notably, SPDIs operate under rigorous regulatory oversight, requiring robust compliance measures, regular audits, and stringent reserve management protocols. Yet, they simultaneously benefit from clear, regulatory frameworks that eliminate much of the uncertainty previously plaguing digital asset ventures. This clarity has significantly reduced operational risk, thereby attracting innovative enterprises and institutional investors eager to engage securely within the digital asset ecosystem.

Further cementing Wyoming's pioneering status is its classification of digital assets within the state's legal code. Wyoming was among the first jurisdictions globally to define digital assets clearly, categorizing them into distinct legal classes, including digital securities, virtual currencies, and digital consumer assets. These clear definitions remove ambiguity, streamline compliance procedures, and enable seamless integration of digital assets into traditional financial operations. Specifically, stablecoins are recognized under Wyoming law as digital assets backed by permissible reserve instruments, such as fiat currencies, government bonds, or commodities. This clear legal recognition significantly enhances stablecoin issuers' credibility, simplifies compliance, and encourages broader market adoption.

Additionally, Wyoming has taken significant steps to establish custodial and fiduciary standards tailored specifically to digital assets. The state's statutes mandate stringent custodial security measures, rigorous auditing procedures, and comprehensive risk management protocols designed to protect consumers and investors. These fiduciary requirements provide critical safeguards against operational risks, cybersecurity threats, and market manipulations, ensuring market stability and reinforcing investor trust. Wyoming's regulatory model thus aligns with the Genius Act's objectives, including transparency, robust consumer protections, and rigorous compliance standards.

Wyoming's regulatory ecosystem is further distinguished by its proactive approach to innovation. The state legislature actively collaborates with industry stakeholders, blockchain innovators, and regulatory experts to continuously refine regulatory frameworks. Regular stakeholder consultations, legislative updates, and agile policymaking processes ensure that Wyoming's regulations remain dynamically responsive to technological advancements and market developments. This agile regulatory approach positions Wyoming strategically as a jurisdiction not merely receptive to innovation but supportive of responsible technological advancements.

The implications of Wyoming's proactive digital asset regulation have been profoundly transformative. The state has attracted numerous blockchain startups, stablecoin issuers, custodial institutions, and technological innovators, significantly contributing to Wyoming's economic diversification and employment growth. Moreover, Wyoming's clear regulatory frameworks have drawn attention from global financial institutions, institutional investors, and digital asset exchanges seeking secure, compliant jurisdictions. This economic influx has bolstered Wyoming's reputation, positioning it strategically as a global leader in digital asset innovation and responsible financial governance.

Wyoming's regulatory clarity has also catalyzed broader national and international regulatory discourse, significantly influencing legislative developments beyond its borders. Legislators drafting the Genius Act cited Wyoming's regulatory model, particularly the SPDI charter and digital asset classifications, as influential precedents informing national regulatory frameworks. Wyoming's success underscores the critical importance of proactive, clear legislative standards as catalysts for economic growth, technological innovation, and financial inclusion, thus validating the Genius Act's strategic vision and regulatory intentions.

Furthermore, Wyoming's proactive stance addresses broader economic and societal considerations, including financial inclusion, ethical stewardship, and responsible governance. By encouraging stablecoin issuance and blockchain innovation within clearly defined regulatory parameters, Wyoming facilitates the development of innovative financial products designed to enhance accessibility, reduce transaction costs, and extend financial services to underserved populations. Wyoming's regulatory model thus aligns strategically with the Genius Act's broader societal objectives, addressing economic stability, consumer protection, and ethical financial innovation.

In summary, Wyoming's pioneering digital asset legislation offers a compelling blueprint aligning with the Genius Act's core regulatory principles, including definitional clarity, rigorous compliance standards, transparency mandates, consumer protections, and proactive innovation support. The state's clear legislative frameworks, proactive regulatory stance, and collaborative policymaking processes have positioned Wyoming strategically as a global leader in digital asset governance, significantly influencing broader regulatory developments. Wyoming's regulatory innovations not only exemplify responsible legislative practices but also offer critical insights and actionable strategies informing ongoing regulatory adaptations and future legislative initiatives, both nationally and globally.

Wyoming's proactive approach has set a clear precedent, demonstrating how regulatory clarity, agile policymaking, and rigorous compliance frameworks can significantly foster responsible innovation, enhance market stability, and position jurisdictions strategically at the forefront of financial and technological advancements.

Analysis of Texas Regulatory Framework and Opportunities

Texas, a state historically synonymous with economic growth and entrepreneurial spirit, has embraced the emerging digital asset ecosystem with an approach distinctly different from Wyoming's proactive legislative frameworks. Characterized by its blend of traditional financial regulation, cautious regulatory adaptations, and market-driven flexibility, Texas offers a unique regulatory landscape full of nuanced opportunities and distinct strategic advantages for businesses navigating the Genius Act's provisions.

Overview of Texas's Regulatory Environment

Unlike Wyoming, Texas has not enacted comprehensive, dedicated legislation governing digital assets and stablecoin issuances. Instead, the state has primarily integrated digital asset regulation within existing financial regulatory structures, relying heavily upon interpretations by state agencies such as the Texas Department of Banking, Texas State Securities Board, and Texas Department of Licensing and Regulation. This integration into traditional frameworks presents both distinct regulatory complexities and notable opportunities for flexibility, potentially advantageous for innovative financial enterprises.

A critical element of Texas's approach is its regulatory philosophy, which emphasizes limited governmental interference, prioritizing market-driven outcomes, self-regulation, and cautious oversight. This philosophy aligns closely with Texas's broader economic principles favoring business innovation, competitive market dynamics, and regulatory restraint. Consequently, digital asset businesses operating in Texas frequently encounter fewer regulatory barriers, potentially lower compliance burdens, and greater operational latitude compared to jurisdictions with more prescriptive legislative frameworks.

Regulatory Interpretations and Agency Guidance

In the absence of comprehensive legislative statutes addressing stablecoins or digital assets, Texas's regulatory landscape relies significantly upon interpretive guidance provided by state regulatory agencies. For instance, the Texas Department of Banking issued interpretive guidance in 2019, clarifying custodial standards and reserve management obligations for financial institutions holding digital assets. This guidance outlined permissible custodial activities, security protocols, fiduciary responsibilities, and reserve verification procedures, offering businesses critical operational clarity and compliance certainty.

Similarly, the Texas State Securities Board has provided extensive interpretive guidance clarifying the classification and regulatory status of digital assets under state securities laws. Leveraging established securities frameworks, Texas regulators evaluate digital assets based on traditional investment contract analyses, closely aligning with the federal SEC's Howey Test criteria. This interpretive consistency significantly enhances regulatory clarity, particularly regarding digital asset issuances classified as securities, thus aligning closely with the Genius Act's definitional clarity objectives.

Licensing Framework and Regulatory Approvals

Texas employs a relatively streamlined licensing process catering to financial institutions and businesses engaging in digital asset transactions. Unlike Wyoming's specialized SPDI charter, Texas primarily utilizes conventional money transmitter licenses and trust company charters adapted to accommodate digital asset custody, stablecoin issuance, and blockchain-based financial services. Texas's licensing framework mandates stringent operational standards, including reserve requirements, transaction transparency, consumer protection obligations, and rigorous AML/CFT compliance protocols, closely aligning with the Genius Act's foundational compliance principles.

Although not tailored exclusively to digital assets, Texas's adaptable licensing frameworks offer significant flexibility and scalability advantages. Businesses benefit from operational latitude, streamlined approval processes, and reduced regulatory complexity relative to jurisdictions imposing specialized digital asset charters. Consequently, innovative startups, fintech enterprises, and established financial institutions frequently find Texas's adaptable licensing processes strategically advantageous, particularly when seeking expedited market entry and operational scalability.

Economic and Market Opportunities

Texas's regulatory landscape leverages the state's broader economic advantages, including extensive financial services infrastructure, robust technology ecosystems, substantial institutional investor presence, and favorable business environments. These broader economic attributes position Texas strategically to capitalize significantly upon digital asset innovations, stablecoin issuances, and blockchain-based financial services facilitated by the Genius Act's regulatory clarity.

The state's robust banking infrastructure, encompassing major national banks, regional financial institutions, and specialized financial services providers, offers critical partnership opportunities beneficial for stablecoin issuers and blockchain innovators. These established financial institutions provide essential custodial services, reserve management capabilities, liquidity provisioning, and transaction settlement infrastructures necessary for compliant stablecoin issuance under Genius Act mandates.

Additionally, Texas's extensive technology ecosystem fosters innovation and strategic collaboration opportunities within blockchain-based financial services. Cities such as Austin, Houston, and Dallas host vibrant fintech clusters, blockchain technology hubs, and extensive entrepreneurial communities supporting innovative ventures. Texas's integration of digital asset services within broader technology innovation clusters significantly enhances strategic opportunities for collaborative partnerships, technological integrations, and operational efficiencies beneficial to Genius Act-compliant businesses.

Regulatory Challenges and Ambiguities

Despite its notable advantages, Texas's reliance upon interpretive agency guidance and traditional licensing frameworks introduces specific regulatory ambiguities and compliance complexities. The absence of comprehensive digital asset-specific statutes creates interpretive uncertainties, operational compliance risks, and potential regulatory inconsistencies. Digital asset businesses operating under Texas's interpretive frameworks frequently encounter nuanced legal complexities, particularly regarding asset classification, reserve verification standards, operational transparency mandates, and fiduciary obligations.

Moreover, Texas's interpretive regulatory landscape demands ongoing proactive engagement with regulatory agencies to ensure continuous compliance and operational clarity. Digital asset businesses require extensive legal resources, regulatory expertise, and strategic compliance infrastructure dedicated to navigating interpretive complexities, agency guidance updates, and regulatory adaptations. These proactive regulatory engagement demands introduce operational overhead, compliance risks, and strategic resource allocations potentially disadvantaging smaller enterprises relative to larger institutions better positioned to absorb regulatory complexities.

Strategic Alignment with the Genius Act

Notwithstanding these regulatory complexities, Texas's market-driven regulatory approach aligns strategically with several core Genius Act objectives, including operational flexibility, compliance adaptability, and market-driven innovation. The state's regulatory restraint, interpretive flexibility, and traditional licensing frameworks facilitate agile compliance adaptations and market responsiveness beneficial under Genius Act mandates.

Texas's strategic market advantages complement Genius Act objectives, including extensive financial infrastructure supporting stablecoin issuance, robust technological ecosystems facilitating innovation, and extensive institutional investor presence catalyzing market adoption. These broader economic attributes position Texas strategically as a compelling jurisdiction capable of leveraging Genius Act regulatory clarity to enhance economic diversification, technological innovation, and financial inclusion beneficial for stakeholders across digital asset ecosystems.

Conclusion on Texas's Regulatory Landscape

In conclusion, Texas offers a distinctly nuanced regulatory environment blending traditional financial frameworks, interpretive flexibility, and market-driven regulatory restraint advantageous for innovative businesses navigating Genius Act compliance. Although characterized by interpretive complexities, operational ambiguities, and compliance demands, Texas's regulatory landscape provides critical operational flexibility, scalability opportunities, and strategic market advantages beneficial under Genius Act provisions.

Texas's integration of digital asset regulations within broader economic ecosystems positions the state strategically to leverage Genius Act clarity to attract stablecoin issuers, blockchain innovators, institutional investors, and financial services providers seeking regulatory adaptability, market responsiveness, and operational scalability. Thus, Texas's regulatory approach complements Genius Act objectives, providing stakeholders with essential flexibility, operational adaptability, and strategic opportunities critical for responsible digital asset innovation, market growth, and economic diversification envisioned by Genius Act legislation.

Comparative Study: Benefits and Drawbacks of Each Jurisdiction

Navigating the regulatory landscape introduced by the Genius Act requires a clear understanding of the strategic implications of selecting a jurisdiction. Wyoming and Texas represent two distinct regulatory frameworks, each offering unique benefits and potential drawbacks. This comparative analysis evaluates each jurisdiction in relation to the Genius Act's core principles—transparency, regulatory clarity, innovation support, economic opportunity, and consumer protection—to provide stakeholders with actionable insights necessary for informed decision-making.

Wyoming: Key Benefits and Strategic Advantages

Wyoming's proactive regulatory approach presents several advantages for businesses engaging in stablecoin issuance and digital asset innovation under the Genius Act. Foremost among these benefits is the regulatory clarity achieved through dedicated digital asset legislation, including clearly defined stablecoin classifications, robust reserve management protocols, stringent custodial standards, and specialized banking charters. This legislative clarity reduces compliance ambiguities, operational risks, and regulatory uncertainty, providing businesses with the legal frameworks necessary to confidently launch innovative financial products.

The Special Purpose Depository Institution (SPDI) charter represents another strategic advantage unique to Wyoming. Designed specifically for digital asset custodianship and transactions, SPDIs offer tailored financial services compliant with Genius Act mandates. This specialized financial institution model facilitates seamless integration between traditional financial infrastructure and emerging blockchain-based services, providing stablecoin issuers with the infrastructure required for efficient reserve management, secure custodial arrangements, and transparent transaction settlement.

Wyoming's regulatory agility further distinguishes it strategically. The state legislature engages in continuous stakeholder dialogue and legislative refinement, ensuring regulatory frameworks remain responsive to technological advancements and market dynamics. This agility positions Wyoming to rapidly adapt to future regulatory developments under the Genius Act, providing businesses with operational flexibility and strategic adaptability critical for long-term innovation and market responsiveness.

Additionally, Wyoming's commitment to innovation and financial inclusion aligns with broader societal objectives emphasized by the Genius Act. The state's supportive environment encourages financial products designed for underserved markets, fostering economic inclusivity, market fairness, and ethical stewardship consistent with the Genius Act's foundational principles.

Wyoming: Potential Drawbacks and Strategic Considerations

Despite its notable advantages, Wyoming's regulatory approach introduces specific strategic limitations relevant to businesses navigating Genius Act compliance. Chief among these is the potential rigidity inherent in specialized regulatory frameworks tailored for digital assets. Businesses operating within Wyoming's legislative parameters may encounter constraints limiting operational flexibility, strategic scalability, and innovative experimentation compared to more adaptable jurisdictions.

Additionally, Wyoming's specialized regulatory environment introduces compliance complexity and associated operational costs higher than jurisdictions employing more generalized licensing frameworks. Compliance with SPDI charter standards, reserve verification protocols, and custodial security mandates requires substantial regulatory expertise, specialized infrastructure investments, and dedicated compliance resources, potentially disadvantaging smaller enterprises relative to established institutions with more substantial compliance capacities.

Moreover, Wyoming's relatively limited financial and technological infrastructure presents strategic challenges relevant to large-scale stablecoin issuances and institutional market integrations. While offering regulatory clarity, Wyoming lacks extensive institutional banking networks, robust investor ecosystems, and large-scale financial service providers necessary for comprehensive market integration advantageous under Genius Act objectives.

Texas: Key Benefits and Strategic Advantages

In contrast, Texas's regulatory landscape offers strategic advantages stemming from its interpretive flexibility, market-driven regulatory restraint, and integration within broader economic ecosystems. Texas employs traditional licensing frameworks adaptable to digital asset operations, providing operational latitude, expedited licensing processes, and reduced regulatory complexity advantageous for innovative financial businesses seeking rapid market entry and operational scalability.

Texas's interpretive regulatory environment facilitates agile compliance adaptations beneficial under Genius Act mandates. Regulatory agencies provide interpretive guidance adaptable to evolving market conditions, technological advancements, and innovative financial products, enabling businesses to strategically align compliance operations with Genius Act regulatory updates and market dynamics.

Strategically, Texas's extensive financial services infrastructure represents another distinct advantage beneficial for stablecoin issuers, blockchain innovators, and financial institutions leveraging Genius Act clarity. Texas hosts extensive banking networks, substantial institutional investor presence, robust financial service providers, and comprehensive transaction settlement systems facilitating efficient operational integration advantageous for large-scale stablecoin issuance and market adoption.

Additionally, Texas's technology innovation ecosystem provides critical strategic opportunities fostering collaboration, technological integration, and operational innovation beneficial under Genius Act provisions. Cities such as Austin, Houston, and Dallas offer vibrant fintech communities, extensive entrepreneurial ecosystems, and innovative technology infrastructures supporting blockchain-based financial services advantageous for stablecoin issuers and digital asset innovators seeking collaborative partnerships, technological synergies, and market integration consistent with Genius Act objectives.

Texas: Potential Drawbacks and Strategic Considerations

Notwithstanding its distinct strategic advantages, Texas's regulatory environment introduces several compliance complexities, interpretive ambiguities, and strategic challenges relevant to Genius Act compliance. Texas's reliance upon interpretive regulatory guidance rather than comprehensive legislative statutes introduces regulatory uncertainties relevant to stablecoin classifications, reserve verification standards, operational transparency mandates, and fiduciary obligations required under Genius Act mandates.

This interpretive regulatory ambiguity necessitates proactive regulatory engagement to ensure ongoing compliance, operational clarity, and interpretive certainty essential under Genius Act compliance obligations. Such regulatory engagement demands substantial legal resources, regulatory expertise, compliance infrastructure investments, and dedicated strategic resources potentially disadvantaging smaller enterprises relative to established financial institutions better positioned to navigate interpretive complexities.

Additionally, Texas's regulatory flexibility presents potential inconsistencies arising from interpretive variability between different regulatory agencies governing digital asset operations within the state. Businesses operating under Texas's interpretive frameworks may encounter strategic challenges arising from divergent regulatory interpretations, overlapping jurisdictional responsibilities, and inconsistent compliance standards potentially complicating operational risk management under Genius Act provisions.

Comparative Strategic Summary and Recommendations

Wyoming and Texas represent two distinctly different regulatory approaches relevant for stakeholders navigating Genius Act compliance. Wyoming's proactive legislative clarity, regulatory definitions, specialized financial institutions, and commitment to agile policymaking offer critical strategic advantages beneficial for innovative digital asset businesses seeking regulatory certainty, operational clarity, and comprehensive compliance frameworks aligned with Genius Act objectives.

Conversely, Texas's interpretive regulatory flexibility, market-driven operational latitude, extensive financial infrastructure, and robust technological ecosystem provide strategic advantages beneficial for businesses seeking regulatory adaptability, rapid market entry, operational scalability, and broad market integration advantageous under Genius Act clarity.

Stakeholders navigating Genius Act compliance must carefully evaluate jurisdictional choices based upon strategic priorities encompassing regulatory clarity, operational flexibility, scalability potential, compliance costs, financial infrastructure integration, technological ecosystem advantages, and broader economic opportunities relevant to business objectives and strategic market positioning under Genius Act mandates.

Ultimately, Wyoming presents optimal strategic alignment for businesses prioritizing regulatory certainty, robust compliance frameworks, specialized financial institutions, and proactive legislative adaptability advantageous under Genius Act compliance obligations. Texas provides strategic advantages beneficial for businesses prioritizing operational adaptability, interpretive flexibility, rapid market entry, broad financial integration, and technological ecosystem synergies strategically beneficial within Genius Act regulatory frameworks.

In conclusion, this comparative analysis equips stakeholders with comprehensive insights necessary for informed strategic decisions essential for leveraging Genius Act regulatory clarity to achieve responsible innovation, market stability, economic diversification, and strategic leadership envisioned by the Genius Act legislation.

Chapter 6: Economic and Ethical Considerations

The Genius Act represents more than a regulatory milestone; it signifies a transformative realignment of digital finance within the broader economic and ethical landscape. As digital assets increasingly influence global financial systems, examining the Genius Act through the lenses of economic stability, consumer protection, and ethical stewardship becomes critical for understanding its comprehensive impact. This chapter meticulously analyzes these considerations, providing a balanced, nuanced perspective designed to guide stakeholders towards informed, responsible decision-making within the new regulatory environment established by the Genius Act.

Economic Stability and Systemic Risk Management

Economic stability is a cornerstone of prosperous societies, and systemic risk management is its essential guardian. The Genius Act addresses these twin concerns by introducing rigorous standards for stablecoin issuers, designed to maintain economic stability and minimize the risk of systemic financial disruptions.

Stablecoins, as digital assets pegged to stable reserve assets such as fiat currency or commodities, present unique economic opportunities alongside inherent risks. These digital instruments have demonstrated substantial potential to streamline payments, reduce transaction costs, and enhance financial inclusion. However, their widespread adoption without stringent regulatory oversight carries pronounced systemic risks. Prior instances of stablecoin instability have demonstrated the capacity for significant economic disruptions, amplifying volatility, undermining investor confidence, and threatening broader financial market stability.

To address these systemic vulnerabilities, the Genius Act introduces stringent reserve management requirements, mandating stablecoin issuers to maintain reserves equivalent to 100% of the issued stablecoins' value. These reserves must comprise high-quality, highly liquid assets such as fiat currencies, government bonds, or precious metals, verified through regular, independent audits. By mandating full reserve backing and rigorous reserve verification procedures, the Genius Act mitigates liquidity risks, solvency concerns, and systemic vulnerabilities, significantly enhancing overall economic stability.

Beyond reserve requirements, the Genius Act establishes transparency mandates designed to enhance economic predictability and reduce systemic risks associated with informational asymmetries. Stablecoin issuers must regularly disclose detailed, transparent information concerning reserve compositions, operational procedures, risk management strategies, and market conditions. These transparency measures empower investors, market participants, and regulators with the critical information necessary for effective economic decision-making, robust risk assessment, and strategic financial management.

Moreover, the Genius Act incorporates comprehensive stress-testing protocols and rigorous capital adequacy standards, designed to assess and enhance stablecoin issuers' resilience under diverse economic conditions. Regularly mandated stress tests evaluate issuer solvency, liquidity resilience, and operational stability under adverse market scenarios, enabling proactive risk mitigation and strategic crisis preparedness. These rigorous economic stability frameworks significantly reduce the probability of issuer insolvencies, reserve inadequacies, or systemic disruptions, thus safeguarding broader economic resilience.

Furthermore, the Genius Act's provisions concerning issuer insolvency resolution and crisis management significantly enhance economic stability by outlining clear, structured protocols for managing operational disruptions or issuer failures. Insolvency procedures define investor protections, market stabilization strategies, and regulatory oversight roles designed to minimize economic disruptions and systemic contagion during issuer insolvencies. These crisis management frameworks provide critical economic safeguards fostering investor confidence, market stability, and broader financial resilience.

An equally important dimension of economic stability addressed by the Genius Act involves comprehensive anti-money laundering (AML) and countering the financing of terrorism (CFT) frameworks. Given digital finance's inherent susceptibility to illicit activities, AML/CFT compliance standards significantly reduce systemic risks associated with illicit financial flows, market abuses, and cross-border regulatory arbitrage. The Genius Act mandates AML/CFT practices—including rigorous customer due diligence, suspicious activity monitoring, and transparent transaction reporting— designed to maintain economic integrity, market transparency, and financial stability.

Collectively, these economic stability measures represent a comprehensive regulatory intervention designed to mitigate systemic risks, enhance market predictability, and foster sustainable economic growth. The Genius Act's rigorous reserve management protocols, transparency mandates, stress-testing frameworks, insolvency resolution strategies, and AML/CFT compliance standards position stablecoin markets for sustained economic resilience and robust systemic stability, safeguarding investor interests and enhancing market confidence.

Consumer Protection and Market Fairness

Consumer protection and market fairness are foundational to any effective financial regulatory framework. Historically, digital asset markets experienced significant challenges arising from insufficient consumer safeguards, informational asymmetries, and widespread market abuses. recognizing these vulnerabilities, the Genius Act integrates rigorous consumer protection standards and market fairness provisions designed to foster trust, transparency, and equitable market practices.

central to the Genius Act's consumer protection framework are stringent transparency obligations. Stablecoin issuers must disclose comprehensive, detailed information concerning operational practices, reserve compositions, risk management procedures, fee structures, and redemption policies. These transparency requirements significantly reduce informational asymmetries, empowering consumers with essential information necessary for informed decision-making, risk assessment, and market engagement.

Another crucial aspect addressed by the Genius Act involves robust liquidity guarantees and redemption assurances for stablecoin holders. Issuers must guarantee timely redemption, maintain sufficient reserve liquidity, and clearly outline redemption procedures, timelines, and associated costs. liquidity assurances significantly reduce consumer liquidity risks, protect stablecoin holders against operational disruptions, and enhance overall market fairness by preventing discriminatory redemption practices.

Additionally, the Genius Act mandates clear, accessible consumer complaint and dispute resolution mechanisms, providing consumers with structured processes for addressing issuer misconduct, operational grievances, or regulatory violations. Complaint management frameworks outline specific timelines, impartial investigation procedures, and transparent outcomes reporting, significantly enhancing consumer trust, issuer accountability, and regulatory oversight.

Investor education and risk communication standards further reinforce the Genius Act's consumer protection objectives. Stablecoin issuers must proactively provide clear, transparent educational resources detailing market risks, operational complexities, reserve management practices, and issuer-specific disclosures. investor education initiatives significantly empower consumers, reduce market misinformation, and enhance overall market fairness by fostering informed consumer engagement.

Moreover, governance transparency standards mandated by the Genius Act significantly enhance issuer accountability and consumer protection by requiring transparent disclosures of board structures, ethical oversight mechanisms, and executive accountability frameworks. governance transparency mitigates potential conflicts of interest, enhances organizational integrity, and fosters responsible market conduct beneficial for consumer trust and market fairness.

addressing market fairness, the Genius Act incorporates rigorous market surveillance protocols targeting market manipulation, trading abuses, and fraudulent practices. Stablecoin issuers and associated market operators must implement sophisticated surveillance systems monitoring real-time transactional volumes, pricing behaviors, and suspicious market activities. manipulation detection and transparent reporting obligations significantly deter market abuses, enhance overall market fairness, and reinforce consumer protections critical for equitable market participation.

Finally, the Genius Act mandates robust consumer compensation mechanisms addressing potential financial losses resulting from issuer insolvencies, operational failures, or regulatory violations. Issuers must maintain dedicated compensation funds, transparently manage claims processes, and ensure timely compensation disbursements safeguarding consumer financial interests. consumer compensation frameworks significantly enhance consumer confidence, issuer accountability, and market fairness essential for sustainable financial market stability.

Collectively, these consumer protection and market fairness provisions represent comprehensive regulatory interventions safeguarding consumer interests, enhancing issuer accountability, and ensuring equitable market practices. By mandating rigorous transparency standards, liquidity assurances, dispute resolution procedures, investor education initiatives, governance transparency, market surveillance protocols, and consumer compensation mechanisms, the Genius Act positions digital asset markets for sustainable consumer trust, market fairness, and regulatory integrity envisioned by responsible financial innovation frameworks.

Ethical Stewardship and Human-Centric Finance

Ethical considerations form the cornerstone of sustainable financial innovation—not merely as guidelines for regulatory compliance, but as integral components fostering long-term growth, public trust, and responsible economic stewardship. Within this context, the Genius Act significantly underscores the importance of ethical governance, ensuring technological advancements in digital finance remain aligned with human-centered principles. By embedding ethical standards deeply into its regulatory frameworks, the Genius Act aims to cultivate a landscape of innovation that benefits society as a whole, rather than prioritizing short-term gains or speculative interests.

At its foundation, ethical stewardship within digital finance involves principles of transparency, accountability, fairness, and respect for human dignity. The Genius Act embraces these ethical dimensions as essential to shaping markets that serve broad societal interests. This deliberate focus on ethics represents a clear shift away from earlier, less-regulated environments, where consumer protection, market integrity, and equitable access were often secondary concerns, sacrificed in the pursuit of rapid market expansion and speculative profitability.

Integrating Ethical Governance within Regulatory Frameworks

The Genius Act integrates ethical governance throughout its regulatory architecture, mandating transparent operations, stringent accountability measures, and rigorous oversight aimed specifically at safeguarding consumer interests, preserving market integrity, and enhancing public confidence. Stablecoin issuers, operating under the provisions of the Act, must adhere to high ethical standards, including thorough disclosures of operational practices, comprehensive reporting on reserve management, and demonstrable mechanisms of accountability to ensure responsible corporate behavior.

Governance transparency provisions require issuers to disclose detailed information regarding their organizational structures, ethical oversight processes, executive accountability frameworks, and risk management protocols. These robust transparency standards strengthen organizational integrity, mitigate conflicts of interest, and foster corporate governance practices aligned with ethical stewardship principles.

Moreover, the Act provides comprehensive whistleblower protections to reinforce ethical accountability by encouraging timely reporting of unethical conduct, regulatory breaches, or operational misconduct. Protecting whistleblowers from retaliation and providing secure reporting channels enhances issuer accountability, promotes ethical compliance, and supports market integrity, thereby strengthening public trust and consumer confidence.

Advancing Financial Inclusion and Equitable Access

Promoting financial inclusion and ensuring equitable access to financial services represent another critical dimension of ethical stewardship emphasized by the Genius Act. Traditional financial systems historically excluded underserved communities due to geographic barriers, infrastructure inadequacies, or socioeconomic disparities. Digital assets and stablecoins, when responsibly regulated, offer considerable potential to overcome these obstacles, extend financial services, reduce transaction costs, and facilitate inclusive economic participation.

The Genius Act incentivizes stablecoin issuers to create innovative financial products specifically tailored to address enduring barriers faced by underserved populations. By offering clearly defined regulatory pathways for responsible innovation, the Act encourages issuers to focus on consumer-oriented financial products designed to enhance financial inclusion, economic accessibility, and equitable market participation aligned with principles of human-centric finance.

For example, stablecoin solutions can significantly lower barriers to entry for populations without traditional bank access, facilitating cost-effective, secure financial transactions even in geographically isolated or economically disadvantaged areas. By reducing transactional frictions, digital assets enable individuals and small businesses to participate meaningfully in the broader economy, directly aligning financial innovation with ethical principles of inclusivity and social equity.

Supporting Responsible Technological Advancement

The ethical stewardship principles outlined in the Genius Act also emphasize responsible technological advancement, highlighting that innovation should prioritize societal welfare and human dignity rather than narrow economic interests. As blockchain technology and digital assets increasingly shape global financial systems, prioritizing human-centric innovation becomes essential to protect public interests, promote equitable participation, and enhance overall societal well-being.

To this end, the Genius Act requires stablecoin issuers to maintain robust technological infrastructures, rigorous cybersecurity standards, and comprehensive operational resilience practices essential for responsible technological advancement. Clearly defined operational standards ensure consumer interests remain safeguarded against cybersecurity threats, technological vulnerabilities, or operational disruptions detrimental to public trust or economic stability.

Specifically, issuers are required to implement stringent cybersecurity measures—including advanced encryption protocols, multi-layered security defenses, regular vulnerability assessments, and incident response planning—to protect sensitive consumer information, transactional integrity, and financial data security. Comprehensive data protection standards reduce systemic vulnerabilities, protect consumer confidentiality, and enhance issuer credibility, thereby promoting ethical market behavior and responsible technological development.

Additionally, issuers must develop, test, and maintain detailed business continuity plans to manage potential operational disruptions, liquidity challenges, or systemic crises effectively. By mandating thorough contingency planning, stress-testing procedures, and crisis response mechanisms, the Genius Act strengthens issuer accountability and operational stability, reinforcing public confidence and ethical market conduct.

Encouraging Ethical Market Behavior and Fairness

The Genius Act seeks to eliminate market abuses and promote ethical market behavior through rigorous market surveillance provisions. Issuers and associated market operators must deploy sophisticated monitoring systems to track transactional volumes, pricing patterns, and suspicious market activities in real time. Comprehensive detection and reporting mechanisms target manipulative practices—such as wash trading, spoofing, pump-and-dump schemes, and coordinated market abuses—ensuring timely identification, deterrence, and corrective action against unethical behaviors.

Robust market surveillance frameworks, coupled with regular integrity reporting and close coordination with regulatory authorities, significantly enhance market transparency, reinforce ethical conduct, and deter fraudulent activities. These measures not only protect consumer interests but also cultivate a fair, equitable, and ethically responsible financial ecosystem beneficial to all market participants.

Incorporating Environmental, Social, and Governance (ESG) Standards

The Genius Act underscores the importance of environmental, social, and governance (ESG) criteria as essential components of ethical stewardship in digital finance. Recognizing the broader societal impacts of financial innovation, the Act mandates ESG compliance standards for stablecoin issuers, compelling businesses to conduct thorough environmental impact assessments, uphold robust social responsibility frameworks, and maintain transparent governance practices.

Stablecoin issuers must transparently disclose ESG performance metrics, including detailed reports on energy consumption, carbon footprints, sustainability initiatives, community engagement practices, financial inclusion strategies, and governance transparency. ESG reporting enhances issuer accountability, fosters ethical corporate behavior, and positions digital finance markets for sustainable, responsible growth aligned with broader societal values.

By embedding ESG considerations within regulatory frameworks, the Genius Act ensures that digital asset markets reflect ethical responsibilities toward environmental sustainability, social equity, and responsible corporate governance. This comprehensive approach positions digital finance innovation not merely as an economic driver but as a powerful force promoting long-term societal well-being and environmental stewardship.

Concluding Reflections on Ethical Stewardship and Human-Centric Finance

In conclusion, the ethical stewardship principles embodied in the Genius Act represent an essential shift towards human-centric financial innovation designed to prioritize transparency, accountability, fairness, financial inclusion, and societal well-being. By embedding ethical governance, responsible technological advancement, consumer protection, market fairness, and ESG standards into its regulatory architecture, the Genius Act positions digital asset markets for sustainable, equitable, and ethically responsible growth.

The Genius Act's comprehensive ethical framework not only addresses immediate market vulnerabilities but also establishes clear precedents for future regulatory approaches aligned with societal values. By fostering responsible innovation, ethical market behavior, and inclusive economic participation, the Genius Act ensures that digital financial advancements benefit the broader public good, safeguard human dignity, and promote equitable societal outcomes.

Chapter 7: Strategic Insights and Future Outlook

In preceding chapters, we meticulously examined the Genius Act's intricate legislative framework, explored its practical implications for businesses, and evaluated critical economic, ethical, and jurisdictional considerations. Having established this foundational knowledge, we now delve deeper into strategic insights, providing policymakers and businesses with actionable recommendations and perspectives on emerging trends and the future regulatory environment. This chapter is strategically structured to guide readers through nuanced policy-making, effective business strategy formulation, and proactive anticipation of future developments.

Strategic Recommendations for Policymakers

The Genius Act has already positioned itself as a landmark in digital asset regulation, reflecting both economic innovation and ethical stewardship. For policymakers, the ongoing challenge lies in balancing regulatory rigor with technological agility and innovation-friendly governance. In this section, we present strategic recommendations designed to assist policymakers in navigating the Act's complexities and harnessing its full potential to foster stable, inclusive, and sustainable financial ecosystems.

1. Foster Continuous Stakeholder Engagement

Policymakers must commit to ongoing engagement with industry stakeholders, including stablecoin issuers, financial institutions, technology innovators, consumer advocacy groups, and academic experts. Structured consultations, regular industry forums, and stakeholder advisory committees provide valuable platforms for dialogue and collaboration. These mechanisms not only facilitate transparency and accountability but also enable regulators to swiftly respond to industry challenges, clarify ambiguities, and refine regulatory guidelines proactively, enhancing overall market confidence and regulatory effectiveness.

2. Invest in Regulatory Technology (RegTech)

Given the complexities inherent in monitoring compliance with stablecoin reserve requirements, transparency obligations, and operational risk standards, policymakers should prioritize investments in regulatory technology (RegTech). Deploying advanced technological solutions such as blockchain-based auditing platforms, automated compliance reporting systems, real-time surveillance software, and sophisticated data analytics significantly enhances regulatory oversight capabilities. These investments streamline regulatory processes, minimize operational costs, and enable agile, efficient enforcement responses, positioning regulators at the forefront of technological and financial innovation.

3. Strengthen Federal-State Coordination Mechanisms

The Genius Act's implementation relies heavily on effective coordination between federal and state regulatory authorities. Policymakers must establish robust interagency coordination frameworks, ensuring seamless information sharing, aligned enforcement standards, and clearly defined jurisdictional responsibilities. Enhancing coordination reduces regulatory overlap, prevents jurisdictional arbitrage, and fosters consistent regulatory interpretations. Policymakers should support interagency committees, regular coordination meetings, and standardized reporting protocols, significantly strengthening regulatory cohesion and market stability.

4. Address Compliance Costs for Small Enterprises

Recognizing the potential compliance burdens faced by small and medium-sized enterprises (SMEs) under the Genius Act, policymakers must strategically introduce tailored support mechanisms. Transitional support measures such as streamlined compliance processes, simplified reporting requirements, regulatory sandboxes, and targeted compliance training programs significantly mitigate compliance burdens for smaller issuers. These initiatives encourage innovation and market entry among SMEs, preserving market diversity and fostering competition without compromising regulatory integrity.

5. Promote Financial Inclusion and Ethical Innovation

Policymakers should explicitly incentivize stablecoin issuers to develop financial products designed specifically for underserved populations. By establishing clear regulatory pathways, policymakers can encourage inclusive innovations, reducing barriers to financial participation, lowering transaction costs, and fostering economic accessibility. Encouraging stablecoin products tailored to meet the needs of low-income communities, rural areas, and historically disadvantaged groups aligns digital finance innovation with broader societal objectives of financial inclusion and ethical stewardship.

6. Enhance International Regulatory Cooperation

Given digital assets' inherently global nature, policymakers must prioritize robust international cooperation frameworks. Establishing mutual recognition agreements (MRAs), harmonizing compliance standards, and engaging actively in international regulatory dialogues significantly enhances cross-border regulatory effectiveness. These strategic cooperation efforts reduce regulatory arbitrage, mitigate systemic risks, and strengthen investor protections globally, positioning the United States as a leader in responsible digital asset governance internationally.

7. Anticipate Technological and Market Evolution

Policymakers must proactively anticipate and strategically adapt to rapid technological advancements and market evolution. Regular assessments of emerging technological trends—such as decentralized finance (DeFi), algorithmic stablecoins, tokenization of assets, and novel blockchain architectures—enable policymakers to stay ahead of potential regulatory challenges. By continuously updating regulatory guidelines and maintaining flexible legislative frameworks, policymakers can effectively manage evolving market risks, foster innovation-friendly environments, and sustainably promote responsible financial innovation.

Business Strategies for Leveraging the Genius Act

The Genius Act marks a pivotal transformation in the digital asset regulatory landscape, creating significant strategic opportunities for businesses. Stablecoin issuers, financial institutions, fintech startups, and blockchain enterprises must thoughtfully and proactively position themselves to leverage this new regulatory environment for long-term success. In this section, we outline key strategies businesses can employ to capitalize on the Genius Act's clarity and regulatory structure, enhancing their competitive advantage and market positioning.

1. Strategic Alignment with Regulatory Clarity

The Genius Act provides unprecedented clarity on stablecoin definitions, asset reserve requirements, transparency obligations, and operational standards. Businesses must integrate this clarity strategically into their core operational models. Clear alignment with regulatory guidelines enables companies to confidently navigate compliance obligations, reducing legal risks, enhancing market credibility, and building trust among investors, regulators, and consumers.

Businesses should prioritize comprehensive internal audits and compliance assessments to ensure alignment with regulatory standards. Detailed evaluations of reserve management practices, governance structures, custodial arrangements, and transparency protocols facilitate robust compliance frameworks. This strategic alignment with regulatory requirements not only mitigates operational risks but also positions companies as market leaders, leveraging compliance as a competitive differentiator.

2. Leveraging Regulatory Technology (RegTech)

To optimize compliance efficiency, businesses should strategically invest in regulatory technology (RegTech). Advanced compliance automation tools, blockchain-based auditing systems, real-time reporting platforms, and automated reserve verification technologies significantly streamline compliance processes. Utilizing these innovative solutions minimizes manual compliance workloads, reduces human errors, and enhances regulatory reporting accuracy.

Businesses that adopt RegTech solutions gain a strategic advantage by freeing internal resources to focus on core business activities, innovation, and market growth. Additionally, proactive investment in regulatory technology positions companies at the forefront of industry best practices, aligning technological innovation with regulatory compliance.

3. Strategic Jurisdictional Selection

As analyzed previously, jurisdictional choices such as Wyoming and Texas offer unique strategic benefits and trade-offs under the Genius Act. Businesses must carefully evaluate jurisdictional options based on their specific strategic priorities—including regulatory certainty, operational flexibility, compliance costs, financial infrastructure integration, and technology ecosystem advantages.

Companies prioritizing regulatory clarity, specialized financial institutions, and proactive governance frameworks may find Wyoming strategically beneficial. Alternatively, businesses seeking operational flexibility, interpretive adaptability, rapid market entry, and extensive financial infrastructure may strategically prefer Texas. A deliberate, informed approach to jurisdictional selection empowers businesses to optimize their regulatory positioning and leverage jurisdictional strengths effectively.

4. Comprehensive Liquidity and Reserve Management Strategies

Rigorous liquidity management and full reserve backing represent critical compliance obligations under the Genius Act. Businesses must strategically implement comprehensive liquidity frameworks, including detailed liquidity forecasting, reserve composition planning, and regular stress-testing scenarios. Developing robust liquidity risk management protocols, clear redemption policies, and detailed contingency funding measures ensures sustainable issuer solvency and market confidence.

Companies should strategically partner with liquidity management experts and financial advisors to gain critical external insights, enhance liquidity management practices, and build robust reserve strategies. This strategic approach enhances issuer resilience, mitigates operational risks, and strengthens overall market positioning.

5. Strategic Collaboration and Partnerships

Strategic partnerships offer significant opportunities for businesses to effectively leverage Genius Act provisions. Stablecoin issuers should actively collaborate with established financial institutions, custodians, technology providers, and compliance specialists. Strategic partnerships with trusted financial institutions facilitate efficient reserve management, custodial arrangements, transaction settlement infrastructure, and liquidity provisioning essential for compliance and operational success.

Businesses should also strategically partner with technology companies and blockchain innovators to leverage advanced technological solutions, enhance compliance capabilities, and foster innovation. Collaboration enables issuers to access broader market networks, expand operational capabilities, and build competitive advantages within the rapidly evolving digital asset landscape.

6. Prioritizing Cybersecurity and Data Protection

Given digital assets' inherent technological complexity, robust cybersecurity frameworks represent a strategic imperative for stablecoin issuers under the Genius Act. Businesses must strategically implement multi-layered cybersecurity defenses, including encryption protocols, secure data storage, real-time threat monitoring, regular vulnerability assessments, and detailed incident response planning.

Companies should strategically engage cybersecurity specialists, conduct frequent security audits, and obtain comprehensive cybersecurity insurance coverage. This proactive approach significantly reduces operational vulnerabilities, enhances issuer credibility, and positions businesses as leaders in data integrity and consumer protection.

7. Transparent Communication and Stakeholder Engagement

Transparent market communication and stakeholder engagement are strategically critical under the Genius Act's transparency obligations. Businesses should strategically establish dedicated investor relations channels, regular transparency reporting practices, and detailed market communication frameworks. Clear, consistent communication fosters investor trust, enhances issuer credibility, and strengthens regulatory compliance.

Strategically allocating internal resources or employing external communication specialists ensures effective stakeholder communication, market transparency, and compliance adherence. Transparent engagement with regulators, investors, and market participants positions businesses strategically as responsible market leaders committed to ethical corporate governance.

8. Integrating Environmental, Social, and Governance (ESG) Practices

The Genius Act explicitly mandates ESG compliance standards, aligning financial innovation with broader societal values. Businesses should strategically integrate comprehensive ESG frameworks into their core operations. Detailed environmental impact assessments, robust social responsibility initiatives, transparent governance practices, and regular ESG performance disclosures significantly enhance market reputation, investor attractiveness, and regulatory alignment.

Businesses strategically prioritizing ESG practices demonstrate corporate responsibility, strengthen consumer trust, and build market differentiation. Leveraging ESG compliance as a strategic advantage positions issuers as leaders in responsible financial innovation and ethical market behavior.

Emerging Trends and the Future Regulatory Environment

As digital finance rapidly evolves, anticipating emerging trends and proactively responding to future regulatory shifts become critical for policymakers, businesses, and market participants. The Genius Act, while comprehensive, represents just one stage in an ongoing regulatory journey. In this section, we explore key emerging trends in digital assets and stablecoins, along with strategic considerations for navigating an evolving regulatory landscape.

1. Decentralized Finance (DeFi) and Regulatory Adaptation

Decentralized finance (DeFi) has rapidly emerged as a transformative trend within digital finance, presenting both significant opportunities and unique regulatory challenges. DeFi applications enable peer-to-peer financial services—including lending, borrowing, trading, and asset management—without traditional intermediaries, relying instead on automated smart contracts and decentralized networks.

As DeFi continues to expand, policymakers must strategically adapt regulatory frameworks to address decentralized architectures, automated transactions, and emerging financial instruments. Future regulatory approaches may focus on setting clear standards for smart contract security, transparent operational practices, consumer protections, and governance mechanisms within decentralized protocols. Policymakers should remain vigilant and proactive, ensuring balanced regulatory oversight that fosters innovation while safeguarding market integrity and consumer interests.

Businesses leveraging DeFi technologies must also anticipate forthcoming regulations, strategically incorporating robust compliance, transparency measures, and risk management protocols into decentralized operations. Those proactively aligning with anticipated DeFi-specific regulatory standards will secure significant market advantages, positioning themselves as responsible leaders within the decentralized finance ecosystem.

2. Central Bank Digital Currencies (CBDCs) and Integration with Stablecoins

Another major trend shaping the digital finance landscape involves central bank digital currencies (CBDCs). Central banks worldwide, including the U.S. Federal Reserve, are actively exploring CBDC implementations to modernize payment systems, enhance monetary policy effectiveness, and promote financial inclusion.

Stablecoin issuers and digital asset enterprises should closely monitor CBDC developments, strategically evaluating potential integration opportunities and competitive implications. Integration between privately issued stablecoins and government-backed CBDCs could offer substantial advantages, enabling seamless interoperability, enhanced liquidity management, and efficient cross-border transactions.

Policymakers must anticipate regulatory adjustments necessary to facilitate potential CBDC and stablecoin integrations. Strategic regulatory frameworks should address interoperability standards, reserve management obligations, consumer protection mechanisms, and technological compatibility. Proactive engagement and collaboration between regulators, central banks, and private stablecoin issuers will be essential for navigating CBDC integration complexities effectively.

3. Asset Tokenization and Expansion of Digital Asset Markets

Asset tokenization, the process of digitally representing real-world assets on blockchain networks, is rapidly gaining prominence. Tokenization offers considerable benefits, including increased liquidity, fractional ownership opportunities, streamlined asset transfers, and enhanced market accessibility.

As asset tokenization expands, stablecoins and digital asset markets will increasingly intersect with traditional asset classes—including securities, commodities, real estate, and intellectual property rights. Policymakers should strategically anticipate and adapt regulatory frameworks to address tokenization-specific considerations such as asset verification standards, custodial arrangements, investor protections, and market transparency obligations.

Businesses must also strategically prepare for increased asset tokenization by developing compliant technological infrastructures, specialized custodial solutions, and innovative tokenization platforms. Strategic investments in asset tokenization capabilities will position businesses advantageously to capitalize on growing market demand, investor interest, and regulatory clarity in this emerging sector.

4. Algorithmic Stablecoins and Regulatory Oversight

Algorithmic stablecoins, which maintain value stability through algorithmically managed supply adjustments rather than explicit asset backing, represent another rapidly emerging trend. Despite significant innovation potential, algorithmic stablecoins carry inherent risks, as evidenced by high-profile collapses and market disruptions.

Policymakers should strategically enhance regulatory oversight to address algorithmic stablecoin vulnerabilities, ensuring operational transparency, robust risk disclosures, and clear stabilization mechanism standards. Future regulatory approaches may focus on rigorous algorithmic audits, transparent reporting obligations, consumer protection safeguards, and mandatory stress-testing protocols specifically tailored to algorithmic stablecoin architectures.

Businesses exploring algorithmic stablecoin issuance must strategically anticipate rigorous future regulatory standards, prioritizing transparency, comprehensive risk management, and operational resilience. Proactive engagement with regulators, investors, and market stakeholders will enable algorithmic stablecoin issuers to effectively navigate regulatory uncertainties, mitigate operational risks, and strategically position themselves within evolving regulatory frameworks.

5. Cross-Border Regulatory Harmonization and Global Standards

As digital asset markets become increasingly globalized, cross-border regulatory harmonization represents a critical future trend. Regulatory divergences between jurisdictions currently pose significant compliance challenges, operational complexities, and investor protection inconsistencies.

Policymakers must proactively pursue international regulatory cooperation, mutual recognition agreements (MRAs), and standardized global compliance frameworks to address cross-border regulatory disparities effectively. Enhanced regulatory harmonization significantly reduces systemic vulnerabilities, minimizes regulatory arbitrage, and fosters global market stability.

Businesses operating internationally must strategically prioritize compliance strategies that anticipate future regulatory harmonization. Leveraging advanced compliance technologies, comprehensive risk management frameworks, and proactive regulatory engagement will enable global businesses to adapt efficiently to harmonized international standards and secure substantial competitive advantages within cross-border digital asset markets.

6. Enhanced ESG Integration and Sustainable Finance Trends

Environmental, Social, and Governance (ESG) considerations are gaining increasing prominence within digital finance, reflecting broader societal expectations of corporate responsibility, environmental stewardship, and social equity. Future regulatory trends will likely expand ESG compliance obligations for stablecoin issuers and digital asset enterprises, reflecting evolving investor demands and global sustainability priorities.

Policymakers should strategically expand ESG compliance standards within regulatory frameworks, requiring detailed environmental impact disclosures, comprehensive social responsibility frameworks, and transparent governance practices for digital asset issuers. Enhanced ESG integration significantly enhances market integrity, investor confidence, and societal alignment.

Businesses should proactively integrate ESG considerations into core operations, strategically investing in sustainability initiatives, social responsibility practices, and governance transparency frameworks. Leveraging ESG compliance as a strategic advantage positions businesses as leaders in responsible financial innovation and sustainable market behavior, securing long-term investor attractiveness and consumer trust.

7. Technological Innovations and Regulatory Agility

Emerging technological innovations—including artificial intelligence (AI), decentralized autonomous organizations (DAOs), quantum computing advancements, and novel blockchain architectures—will significantly shape future regulatory landscapes. Policymakers must strategically anticipate technological disruptions, ensuring regulatory frameworks maintain adaptability, responsiveness, and future-proof flexibility.

Regular technological assessments, industry dialogues, and adaptive regulatory strategies enable policymakers to remain agile amid rapid technological changes. Regulatory agility fosters responsible innovation, market stability, and consumer protections while minimizing regulatory obsolescence or technological incompatibilities.

Businesses must strategically monitor technological advancements, proactively adapting compliance practices, technological infrastructures, and operational frameworks to emerging technologies. Proactive technological foresight and strategic regulatory alignment enable businesses to leverage future innovations effectively, reduce operational risks, and secure long-term market positioning advantages.

Concluding Reflections and Strategic Outlook

The Genius Act represents a foundational milestone, yet the regulatory landscape of digital finance remains dynamic, continuously evolving alongside technological advancements, market developments, and societal expectations. Policymakers, businesses, and stakeholders must strategically embrace forward-looking perspectives, proactively anticipating future regulatory trends and market disruptions.

Proactive stakeholder engagement, international regulatory harmonization, technological agility, ESG integration, and responsible market practices will remain strategically critical for sustained success within evolving digital asset markets. By embracing strategic foresight, proactive compliance practices, ethical innovation frameworks, and responsive regulatory strategies, stakeholders can confidently navigate future uncertainties, leverage regulatory clarity effectively, and position themselves advantageously at the forefront of responsible, sustainable digital finance innovation.

As digital asset markets mature, strategic insights, proactive regulatory engagement, and informed business strategies become increasingly essential. The Genius Act provides comprehensive foundational clarity, yet strategic adaptability, technological foresight, and ethical stewardship remain continuously critical for long-term market leadership, societal trust, and economic sustainability.

Conclusion: Synthesizing Knowledge and Action Steps

The Genius Act emerges as a landmark legislative achievement, signaling a transformative era in the landscape of digital finance regulation. Throughout this comprehensive exploration, we have systematically unpacked the intricate dimensions of this landmark legislation, examining its historical roots, strategic vision, economic implications, ethical considerations, and jurisdictional complexities. As we conclude our detailed analysis, it is essential to reflect on key insights, actionable steps, and the broader implications for the future of digital finance.

Key Takeaways and Summary of Insights

At the heart of the Genius Act lies a strategic vision aimed at harnessing the transformative potential of digital assets, particularly stablecoins, while safeguarding economic stability, protecting consumers, and ensuring ethical financial practices. The Act's inception was driven by a clear recognition of the inadequacies inherent in previous fragmented regulatory frameworks, which frequently produced ambiguity, systemic risks, and inhibited responsible innovation. Thus, the Genius Act represents a deliberate and carefully crafted response, designed to address these vulnerabilities comprehensively and proactively.

One of the most significant insights derived from this exploration is the Act's explicit approach to stablecoin regulation. By clearly defining stablecoins and distinguishing them from volatile cryptocurrencies, the Act provides foundational clarity crucial for robust compliance, investor confidence, and market integrity. Stablecoin issuers are now equipped with precise regulatory guidelines governing reserve requirements, operational transparency, and fiduciary obligations, significantly mitigating previous systemic risks associated with inadequate asset backing and opaque operational practices.

The Genius Act's emphasis on transparency emerges as another critical insight, explicitly recognized as essential for reducing informational asymmetry, fostering investor trust, and enhancing market stability. Issuers must adhere to stringent disclosure standards, detailing reserve asset compositions, liquidity conditions, operational practices, and associated risks. This transparency empowers investors and regulators alike, facilitating informed decision-making and proactive risk management practices.

A vital component of the Genius Act's success lies in its carefully structured compliance and enforcement mechanisms. These frameworks ensure issuer accountability through mandatory compliance programs, robust oversight procedures, and clearly defined enforcement penalties. Regulatory authorities are now empowered with extensive investigatory capabilities, enabling swift and targeted responses to potential compliance violations or market abuses. Such rigorous enforcement standards significantly strengthen market integrity and consumer protections, instilling widespread confidence in digital finance markets.

In terms of jurisdictional analysis, the contrasting regulatory landscapes of Wyoming and Texas offered valuable insights into the strategic considerations shaping jurisdictional selection. Wyoming's proactive and specialized regulatory frameworks provide unparalleled clarity and explicit support for digital asset ventures. Conversely, Texas's more flexible, market-driven interpretive approach offers significant scalability advantages and operational adaptability. Stakeholders must carefully weigh these jurisdictional trade-offs based on their unique strategic priorities and market objectives.

The economic and ethical considerations outlined by the Genius Act underscore a profound commitment to safeguarding systemic stability, promoting market fairness, and embedding ethical stewardship into digital finance practices. The Act strategically aligns economic stability objectives with comprehensive reserve management protocols, transparent disclosure obligations, robust crisis management frameworks, and explicit AML/CFT compliance standards. Simultaneously, ethical stewardship and human-centric finance principles guide regulatory frameworks, prioritizing financial inclusion, responsible technological innovation, and comprehensive ESG integration.

Finally, the future-oriented perspective inherent in the Genius Act acknowledges that the regulatory landscape will continue evolving alongside technological advancements and global market trends. Stakeholders must proactively anticipate future developments—such as decentralized finance, central bank digital currencies, asset tokenization, algorithmic stablecoins, international regulatory harmonization, and enhanced ESG compliance—and strategically position themselves to leverage these emerging opportunities responsibly.

Actionable Steps for Policymakers and Entrepreneurs

To fully realize the potential of the Genius Act, policymakers and entrepreneurs must strategically undertake a series of targeted, actionable steps designed to maximize compliance effectiveness, enhance market competitiveness, and foster sustainable digital finance innovation.

Actionable Steps for Policymakers:

1. Strengthen Regulatory Collaboration:
Establish structured, ongoing stakeholder dialogues and interagency coordination mechanisms to enhance regulatory responsiveness, clarify compliance obligations, and adapt to evolving market developments proactively.

2. Invest Strategically in Regulatory Technology:
Deploy advanced RegTech solutions—such as blockchain-based reserve verification platforms, automated compliance reporting systems, and real-time market surveillance tools—to streamline oversight, reduce operational costs, and enhance enforcement efficiency.

3. Address Compliance Costs for Small Enterprises:
Implement tailored regulatory support mechanisms for SMEs, including simplified compliance frameworks, transitional regulatory sandboxes, targeted compliance training initiatives, and streamlined licensing processes to foster market diversity and responsible innovation.

4. Expand Financial Inclusion Initiatives:
Incentivize stablecoin issuers to develop inclusive financial products specifically targeting underserved communities, reducing barriers to financial participation, lowering transaction costs, and facilitating economic accessibility and equitable market participation.

5. Prioritize International Regulatory Cooperation:
Establish comprehensive international coordination frameworks, mutual recognition agreements (MRAs), and standardized global compliance standards to reduce cross-border regulatory arbitrage, enhance global market stability, and strengthen investor protections.

6. Foster Regulatory Agility and Technological Foresight:
Regularly conduct technological assessments and proactively update regulatory guidelines, ensuring adaptability, responsiveness, and future-proof flexibility amidst rapidly evolving technological advancements.

7. Deepen ESG Integration:
Expand ESG compliance obligations for stablecoin issuers, mandating detailed environmental impact disclosures, robust social responsibility frameworks, and transparent governance practices to align digital finance with broader societal sustainability expectations.

Actionable Steps for Entrepreneurs and Businesses:

1. Proactive Regulatory Alignment:
Conduct thorough compliance audits and strategically align operational practices, reserve management protocols, transparency standards, and risk management frameworks explicitly with Genius Act requirements.

2. Strategic Jurisdictional Selection:
Carefully evaluate jurisdictional advantages—such as Wyoming's regulatory clarity versus Texas's operational flexibility—selecting jurisdictions strategically aligned with business objectives, regulatory preferences, and scalability priorities.

3. Comprehensive Liquidity Management:
Implement robust liquidity forecasting models, strategic reserve composition planning, and regular liquidity stress-testing scenarios to ensure sustained issuer solvency, market resilience, and compliance effectiveness.

4. Invest in Regulatory Technology Solutions:
Strategically deploy advanced RegTech solutions to automate compliance processes, reduce manual workloads, enhance reporting accuracy, and proactively manage operational risks and compliance complexities.

5. Strengthen Cybersecurity Frameworks:
Establish multi-layered cybersecurity protocols—including encryption technologies, secure data management, regular vulnerability assessments, and explicit incident response plans—to proactively safeguard data integrity, operational resilience, and consumer trust.

6. Foster Transparent Stakeholder Communication:
Strategically allocate resources to establish dedicated transparency and investor communication channels, providing regular, detailed operational reports, compliance certifications, and market updates to enhance investor confidence and regulatory credibility.

7. Integrate Comprehensive ESG Practices:
Explicitly incorporate ESG considerations into core operational frameworks, conducting detailed environmental

assessments, robust social responsibility initiatives, and transparent governance practices to strengthen market reputation, investor attractiveness, and ethical corporate stewardship.

Final Reflections on the Future of Digital Finance

The Genius Act signifies more than just a regulatory milestone—it represents a visionary commitment to fostering responsible, sustainable, and inclusive financial innovation. As digital asset markets evolve, the Act sets a precedent for future legislative efforts, underscoring the importance of regulatory clarity, proactive oversight, ethical governance, and adaptive market frameworks.

Looking forward, the trajectory of digital finance regulation will undoubtedly continue to unfold dynamically. The emergence of decentralized finance, central bank digital currencies, asset tokenization, and algorithmic stablecoins necessitates agile regulatory adaptation, strategic foresight, and proactive stakeholder collaboration. Policymakers must maintain regulatory agility, continuously assessing emerging market trends and technological advancements to ensure responsible governance and systemic stability.

Entrepreneurs and market participants must similarly remain proactive, strategically anticipating future regulatory developments, investing in advanced technological solutions, and prioritizing ethical market practices. Businesses proactively embracing regulatory clarity, compliance excellence, technological innovation, and ESG integration will be strategically positioned for sustainable market leadership and long-term competitive advantage.

Ultimately, the Genius Act serves not merely as regulatory guidance but as an invitation for continued dialogue, collaboration, and strategic innovation within digital finance. As stakeholders collectively navigate evolving market landscapes and technological transformations, maintaining proactive engagement, ethical stewardship, and strategic adaptability will remain critically essential.

By embracing these strategic imperatives, policymakers, entrepreneurs, and market participants can confidently leverage the transformative potential of digital finance—fostering economic resilience, ethical innovation, and inclusive financial ecosystems for generations to come.

Appendices

Appendix A: Quick-reference Summary of Genius Act Provisions

The Genius Act represents a significant milestone in digital asset regulation, particularly focusing on stablecoins. Below is a concise, quick-reference summary outlining the Act's critical provisions to facilitate rapid consultation and informed decision-making by policymakers, entrepreneurs, and compliance officers:

1. Definitions and Classifications:

- **Stablecoin Definition:** Digital assets designed to maintain stable value through reserves such as fiat currencies, commodities, government bonds, or other approved financial instruments.
- **Asset-Backed vs. Algorithmic Stablecoins:** Differentiates clearly between stablecoins backed by tangible audited reserves and those stabilized by algorithmic mechanisms.
- **Systemically Significant Stablecoins:** Identifies stablecoins whose scale could impact broader economic stability, subjecting them to enhanced oversight and stringent standards.

2. Asset Reserve Requirements:

- **Full Reserve Backing:** Mandates issuers to maintain reserves equal to 100% of issued stablecoins.

- **Qualifying Reserve Assets:** Includes fiat currencies, government bonds, highly liquid securities, and precious metals.
- **Custodial Standards:** Reserves must be securely held by qualified custodians with regular independent audits and publicly available audit reports.

3. Transparency and Disclosure Obligations:

- **Reserve Transparency:** Regular public disclosure of detailed reserve compositions, liquidity conditions, and custodial arrangements.
- **Operational Transparency:** Clear public documentation of operational procedures, pricing mechanisms, redemption policies, and governance structures.
- **Risk Disclosures:** Mandatory disclosure of liquidity, custodial, market, and operational risks.

4. Compliance and Enforcement:

- **Mandatory Compliance Programs:** Comprehensive internal controls, defined policies, and regular training.
- **Regulatory Oversight and Investigations:** Extensive authority for regulators to conduct inspections, investigations, and enforce penalties for non-compliance.
- **Penalty Provisions:** Clearly defined consequences for regulatory violations, including substantial fines and potential operational restrictions.

5. Governance and Accountability:

- **Board Accountability:** Explicit responsibility for compliance oversight and fiduciary duties by directors and governing bodies.
- **Internal Audit Requirements:** Regular internal audits of compliance effectiveness, operational integrity, and risk management procedures.
- **Whistleblower Protections:** Comprehensive protection mechanisms for individuals reporting misconduct or regulatory breaches.

6. Jurisdictional Clarification:

- **Federal Primacy:** Centralized federal oversight by agencies such as SEC and CFTC.
- **State Roles and Coordination:** State licensing and consumer protection roles clearly defined and harmonized with federal standards.
- **Interagency Coordination:** Formalized communication and cooperation mechanisms between federal and state regulators.

7. Operational Risk Management:

- **Technological Infrastructure:** Robust and resilient technology standards for issuer operational continuity.
- **Cybersecurity Requirements:** Comprehensive cybersecurity measures, including encryption, regular vulnerability assessments, and secure incident response planning.

- **Business Continuity Planning:** Mandatory detailed contingency planning to address operational disruptions and crisis scenarios.

8. Consumer Protection Measures:

- **Liquidity Guarantees:** Assurances of timely redemption and transparent redemption policies for stablecoin holders.
- **Complaint and Dispute Resolution:** Clear and accessible consumer redress processes for addressing operational grievances or issuer misconduct.
- **Investor Education:** Mandatory investor communication and education to ensure informed consumer participation.

9. Environmental, Social, and Governance (ESG) Standards:

- **Mandatory ESG Compliance:** Issuers required to disclose detailed ESG performance metrics, reflecting environmental impact, social responsibility, and governance transparency.

Appendix B: Resource Guide and Recommended Readings

To further deepen understanding and facilitate comprehensive exploration of the Genius Act and related topics in digital finance regulation, the following curated resource guide provides essential readings and authoritative sources beneficial for policymakers, entrepreneurs, technologists, and financial professionals:

Authoritative Texts and Reports:

- **"The Future of Money: Digital Currency" by Eswar S. Prasad (2021)** – Comprehensive exploration of digital currencies and their transformative implications on global financial systems.
- **"Blockchain Revolution: How the Technology Behind Bitcoin is Changing Money, Business, and the World" by Don Tapscott and Alex Tapscott (2016)** – Seminal text outlining blockchain technology's foundational principles and future potential.
- **"Cryptoassets: The Innovative Investor's Guide to Bitcoin and Beyond" by Chris Burniske and Jack Tatar (2017)** – Analysis of digital assets' economic and investment implications.

Legislative and Regulatory Resources:

- **United States Congress – The Genius Act (S. 394/S. 1582)** (congress.gov) – Official legislative text and congressional documentation for detailed statutory analysis.

- **Wyoming Digital Asset Legislative Framework** – Comprehensive details on Wyoming's regulatory approach, offering valuable jurisdictional insights and strategic frameworks relevant for stablecoin issuers.
- **Texas Department of Banking Interpretive Guidance on Digital Assets (2019)** – Regulatory insights into Texas's interpretive approach, highlighting flexible licensing frameworks and compliance practices.

Regulatory Technology and Compliance:

- **"Regulatory Technology Handbook: How to Leverage RegTech for Efficient Compliance Management" by Janos Barberis and Douglas W. Arner (2019)** – Comprehensive guide on RegTech implementation and its strategic value in compliance optimization.
- **RegTech Association (regtech.org)** – Professional association providing updated resources, best practices, and global developments in regulatory technology solutions.

International Regulatory Context:

- **European Union – Markets in Crypto-assets (MiCA) Regulation** – Comprehensive regulatory standards established by the European Union, providing comparative analysis benchmarks relevant to Genius Act alignment.
- **Monetary Authority of Singapore (MAS) – Digital Token Guidelines** – Regulatory standards and compliance frameworks from Singapore, illustrating

effective risk management and consumer protection approaches.

ESG and Ethical Considerations:

- **"The Sustainable Investing Handbook" by Alan S. Gutterman (2021)** – Insights into ESG integration, investor expectations, and sustainability practices in contemporary financial markets.
- **Principles for Responsible Investment (PRI)** (unpri.org) – Global initiative providing resources and guidelines for effective ESG integration in financial operations.

Strategic and Future-oriented Insights:

- **"The Token Economy: How the Web3 Reinvents the Internet" by Shermin Voshmgir (2020)** – Exploration of tokenization trends, decentralized finance developments, and strategic implications for regulatory frameworks.
- **"Central Bank Digital Currencies: Policy Issues and Challenges" by Tobias Adrian and Tommaso Mancini-Griffoli (IMF, 2021)** – In-depth analysis of central bank digital currencies, strategic implications, and future regulatory adaptations.

Industry Periodicals and Journals:

- **Journal of Financial Regulation and Compliance** – Regular updates on regulatory developments, compliance best practices, and in-depth analysis of financial legislation including stablecoins.
- **Blockchain and Cryptocurrency Journal** – Peer-reviewed research and analysis on blockchain

technology applications, market trends, and regulatory implications.

Websites and Databases:

- **Financial Stability Oversight Council (FSOC)** (fsoc.gov) – Official resource providing regular assessments of systemic risk and financial stability implications relevant to digital assets.
- **U.S. Securities and Exchange Commission (SEC)** (sec.gov) – Comprehensive resources on securities regulation, compliance standards, and regular updates relevant to digital asset issuers.

This curated resource guide ensures stakeholders possess authoritative references and relevant materials for comprehensive understanding, effective compliance alignment, and strategic engagement with the Genius Act's evolving regulatory landscape.

Made in United States
Orlando, FL
22 September 2025

65259106R00089